KT-214-310

MILITARIA

Frontispiece:

Many musical compositions of the 1850s and 60s were given a military flavour. This example of 1856, published by M. Distin, is typical.

MILITARY QUADRILLE,

PERFORMED WITH GREAT SUCCESS AT THE

PROMENADE CONCERTS,

MILITARIA

by
Frederick Wilkinson

WARD LOCK & CO. LIMITED
London and Sydney

7063 1011 X

TO
TERESA

Printed in Great Britain
by Richard Clay (The Chaucer Press), Ltd.,
Bungay Suffolk
Set in Perpetua Monotype Series 239

CONTENTS

Acknowledgments

Once again I most gladly record my gratitude to the many friends, collectors and dealers who gave so generously of their time and knowledge in helping me prepare this book. Many went to a great deal of trouble and lent items from their collections and I am most especially in their debt.

Particular thanks are also due to Ken Wynn who supplied numerous out-of-print books and similar material. As always Gerald Mungeam helped in so many ways. Ted Holmes and Paul Forrester took most of the photographs and deserve special thanks.

The publishers and I are grateful to the following for permission to reproduce the photographs on the following pages:

R. Amos 104 (top), 140, 144, 189, 204 (bottom), 205, 206, 207; Morris Angel & Son Ltd 146, 168, 173, 177; L. Archer 28, 106, 115, 116, 117, 118, 119, 120 (lower), 121 (lower), 123 (lower), 124, 125 (top); G. Bennett 141, 143, 161, 164, 165, 175, 178; H. Blackmore 78 (bottom); Carreras Ltd 100, 101; P. Dale Ltd 126, 136, 139 (top); Mr & Mrs R. Gregory 176 (top), 193, 194, 196, 197, 198, 201, 202, 204 (top); D. H. S. Gyngell 200 (top); J. B. Hayward, Piccadilly 49, 50, 51, 52, 53; J. Luck 169; G. Mungeam 21, 36 (bottom), 91; National Army Museum 58, 61, 69, 70, 71, 72, 73, 180; Parker Gallery 25, 29, 34 (top), 35, 37, 38; John Player & Sons Ltd 98, 99; F. Stephens 54, 55,

56, 57, 190, 203 (bottom); Trustees of the Wallace Collection 157, 158, 159, 160, 162, 163; Dr P. Willis (Jacket illustration); Winchester Gun Museum 138 (centre and lower), 139 (top); P. Woollacott 124, 125. The list of Regimental Museums (pages 233–249) is reproduced by courtesy of *Soldier* (H.M.S.O.).

INTRODUCTION

"The profession of a soldier is allowed to be famous and honourable amongst men that kings and emperors have been enrolled in it" (Capt. J. S., *Military Discipline*, 1685).

"There is no calling under Heaven which hath not beene by malignant spirites traduced and vilified; and amongst all none hath been more disgraced with opprobrious language than that of a soldier" (D. Lupton, *Treatise against the Pike*, 1642).

Two approaches which fairly well typify reactions to the soldier common to almost every land, every period and every race. On the one hand, the military life is extolled as being honourable, gay and gallant – at the other extreme no vilification is too great, and all who therein partake are worse than thieves or murderers. Attitudes prevailing at any one time are usually determined by the need for military protection – in times of great danger the army receives most praise and support.

Unfortunately war has proved to be one, if not the most, absorbing preoccupation of man throughout the whole of his history. Aggression is part of his makeup, and no matter how strongly he seeks to suppress it, the latent spark will, in certain circumstances, flare up. There can never be any real justification for war – at best it is the lesser of two evils – but it has played so great a part in man's story that its effects cannot be ignored.

War has always stimulated opposites – at one and the same time it is dirty, degrading, humiliating, stupid and pointless, and yet evokes acts of supreme heroism, selfless devotion, glamour and even joy.

For most of us it is a pastime to be enjoyed vicariously – it can be exciting and desirable when there are no dangers or personal discomforts involved. There are few more enjoyable and less harmful ways of indulging in the delights of the military life than that of collecting militaria. Such collecting does not, in any way, imply approval of war, but only its acceptance as an unfortunate and deplorable fact of history and one facet of human endeavour that has led to the production of some articles of great interest.

By definition the term "militaria" should be used only for things military (Latin: *miles* – a soldier), but in collecting circles it has gradually been extended to embrace almost anything connected with the armed services. Perhaps it is time that collectors coined a new omnibus phrase, but at present "militaria" suffices and is understood by collectors. In this book the term has been interpreted in a very broad sense embracing some aspects which the purists may feel should not have been included. In justification, if any be needed, it may be said that the term has always been so casually applied that, in fact, there has been no unqualified extension.

Some information has been included on police items, since a growing number of collectors are becoming interested in this rather specialised field.

Militaria offers a number of attractions to be found in few other fields of collecting. For the enthusiast with limited resources it still offers good opportunities to build up an extensive collection of small items at a very modest cost. Excluding such things as the rarer medals, early headdresses, uniforms and weapons, there are still a great many groups of objects which can be acquired for modest sums.

It is a hobby which can be active or passive. As in every sphere of collecting, there is so much to be learned that the collector will find ample opportunities for research. In some cases he will often find himself to be the pioneer in that particular field of study. For the more active enthusiasts model soldiers afford an outlet which can be creative as well as exact. Similarly, war games – one of the newest branches of military hobbies – can be very stimulating.

This book aims to do no more than introduce the reader to some of the more popular branches of militaria, but, of course, this is not to say that there are not other fields to be explored. Military collectors tend to specialise to an extraordinary degree, so that the opportunities are legion.

Each chapter covers one main topic, and as in the author's previous volumes the emphasis has been placed on the more usual and available items rather than the outstanding, rare or unique pieces.

Infantry of the Continental Army, reproduced from a print by H. Ogden. The uniforms are very similar to those of the British Army; the muskets are French in style. Original series issued *c.* 1900.

I

MILITARY PRINTS

Military prints have for long been popular with a variety of collectors, some of whom are attracted by their military interests, others by their artistic qualities. Apart from original water-colours and oils, often done by an enthusiastic and capable amateur, the great majority of these prints are, strictly speaking, book plates. Coloured plates have been produced since the late 18th century, but early examples were, of course, largely hand coloured, and the use of printed colour dates only from the 1840s. Coloured plates were found to sell well, and print sellers very soon realised that they could expect a better return from the sale of individual plates than for the entire book, and it became common practice to remove the plates and discard the text. Many of the illustrations were based on famous engravings and paintings, and the print collector soon comes to recognise the variations on a theme.

Printed books of the 16th century contain engravings which are, on the whole, rather crude and lacking in finesse. It was normal practice to depict all characters in contemporary costume, thus it is not at all unusual to see Romans and Greeks in full-plate armour. However, it is this very feature that makes some of them useful for identifying arms and armour of the 16th century. These prints are not uncommon, and sometimes the whole page, sometimes simply a portion, may turn up.

During the first half of the 17th century a large number of military works were published. Some were histories of wars, but others were designed to instruct the reader on how to become a proficient soldier. One which was first published in 1607 was illustrated by Jacob De Gheyn and showed in a series of large, bold engravings how to handle the musket, caliver and pike. This series of engravings was frequently copied, and variations will be found in many later books. Some of these copies have been coloured by hand at a later date, but they were originally black-and-white engravings.

Many 17th-century authors sought to give an overall coverage on the subject of war. Volumes such as Robert Ward's *Animadversions of War* published in 1639 are illustrated with a large number of wood cuts of many diverse military items, such as mines, fortifications and incendiary devices, and again pages are frequently to be found either mounted or loose. In addition to such instruction books there were many other publications which detailed sieges and battles, especially those of the Low-countries or Netherlands. Such engravings show little detail, for the plates are often crowded and the figures are correspondingly small, but a certain amount of information on tactics, strategy and battle formations can be elicited. As far as uniforms are concerned, such engravings are of very limited use, as all colour is lacking.

The advent of the 18th century in no way abated this flood of books and instruction manuals on military subjects, frequently illustrated by a series of plates. During the latter part of the century there appeared a number of books whose plates have since become familiar to all military enthusiasts. Francis Grose has become well known for his work on the history of the British Army published in 1801. In this work he included a number of plates, many based on earlier works by De Gheyn and others. It must be said that many prints in this book dealing with the uniform of earlier periods are far more imaginative than accurate.

With the outbreak of the Napoleonic Wars there was a tremendous increase in the number of volunteer groups, and many of the well-known artists began to cater for the demand for military subjects. One of the best known of such books is *The Loyal Volunteers of London and its Environs*, published by Ackermann, with eighty-seven plates executed by the famous artist T. Rowlandson. Each large coloured plate shows one movement of the arms drill being demonstrated by a member of one specific volunteer company. Opposite each plate is a brief history of the group with details of the officers and the uniform and date of formation. The complete volume, of course, is extremely expensive, but single plates do turn up and are much sought after.

Instruction books illustrated the use of the sabre and the bayonet, and again single plates may be found, often mounted and glazed. Another volume which has provided a fair number of single plates is *Historic and Military Anecdotes*, by E. Orme, published in 1819, which contained twenty aquatints depicting stirring deeds of the Napoleonic Wars.

Military interest remained high, and during the period from 1830 to 1850 a large number of military books were published, including a very well-known set by R. Cannon, who wrote and edited a series of small regimental histories. These volumes, some sixty-eight in all, contained coloured plates showing the uniform of the regiment, and again they are often found separate. However, not all military prints were produced for book illustration, and a number of single plates or portfolios were issued during the first half of the 19th century, including a fine set by an Irish painter, Michael Angelo Hayes. These prints were frequently copied and modified by later artists, as for example, the set of lithographs issued in 1846 by a man named Walker.

During the century the flood of military books continued unabated, increasing during the middle of the century, when interest and concern was aroused by the Indian Mutiny, Crimean War and the sudden realisation that England's military machine had fallen into disrepair. Many volunteer groups, particularly rifle regiments, were once again formed, and a growing demand stimulated writers to produce military histories, do-it-yourself shooting guides and similar works illustrated with simple line drawings. There was at the same period an increase in newspaper and magazine coverage, and journals such as *The Illustrated London News* often carried double-page coverage of battles as well as reporting on new equipment and changing styles of uniform. Similarly, when *The Boy's Own Paper* began publishing in 1879 it carried an occasional feature in the form of a double-page colour plate dealing with some aspect of military life or the history of the armed forces with such titles as "Our British Soldier" (1880), "Our National Defenders" (1881). "Our British Volunteers" (1881), "Our Military Bands" (1885) and "Colours of the British Army" (1888). Since these were contemporary records, the details can usually be accepted as being accurate.

It is to Regimental Histories that most of the colour plates of the time owe their origin, and among the illustrators two names stand out – Richard Simkin (1850–1926) and Captain H. Oakes-Jones. Both these talented artists produced a large number of attractively coloured and accurately presented plates of uniforms and events in the history of the British Army. It must be recorded with sorrow that when Captain Oakes-Jones died, not very long ago, a certain amount of his personal material was disposed of, and the extent of the loss may never be known. Artists such as Simkin were influenced by many of the late Victorian artists who were inspired to produce great battle paintings. One such well-known artist was Richard Caton Woodville, who painted spirited battle pictures of Blenheim and Candahar. Many of these paintings were copied and engravings produced in quantity. William Heysham Overend was well known for his naval combat drawings while Simpson was renowned for his paintings of the Crimean War. It is not out of place to mention here that the advent of the camera in the middle of the 19th century opened up a new field for collectors. Much of our information on some of the less-well-known regiments and actions can be deduced from photographs taken during this period. No opportunity should be neglected by the enthusiast to check over regimental magazines, newspapers, particularly local ones, for photographs of this period. Quality is sometimes rather poor, but even so the amount of useful information to be gained from such material cannot be over-estimated. Unfortunately a great deal of this material is lost, for such photographs are usually personal mementoes, and upon the death of the owner may be thrown out.

At the same time, on the Continent, particularly France, there was a similar military interest, and one particularly useful series of small books or magazines are the *Cahiers d'Enseignement Illustré*. This series began publication in 1884, and the great majority of issues contain at least one article of the army or military life. The eighty-seven separate issues cover the following armies in greater or lesser detail: French, German, English, Swedish, Spanish, Austrian, Italian, Danish and Norwegian. Other issues contain articles on life in barracks, on manoeuvres, or deal with associated topics such as the Sapeur-Pompier, the firemen, and some of the French Colonies. Not all the illustrations are done by the same man, and most were drawn by Armand Danaresq and Marius Roy. Once again, as they are contemporary records, their accuracy may be accepted.

The 20th century, with the great advances in printing and colour printing, saw a flood of regimental histories, many of which were illustrated, often by the author, with plates culled from official records or copied from paintings or portraits in the possession of families long associated with the particular regiment. One rather unfortunate off-shoot of the technological advances has been the increased use of new methods in the production of what can only be called facsimiles. Many of these are so good that they almost defy detection, especially when sold glazed. In theory one should be able to ask for the back to be removed from the frame, but it would be a very brave collector who did so! One small indication may be seen in the evenness of print, for whereas a true engraving, even the best of them, tends to be variable in density and distribution, the new electronic processes tend to produce an overall even effect not to be found in the originals.

In the United States H. G. Ogden did some forty-four fine-quality colour plates depicting the uniforms of the American Army from the War of Independence until the end of the 19th century, and this is one series that has been reprinted. Another American series produced in 1893 was that entitled *Military Types of U.S. Militia and National Guard Past and Present*, a series of colour plates by L. Prang & Co., which is painted by Aug. Tholey, although so far these do not seem to have been re-issued.

There are probably two indispensable books for the print collector, and they are White's *Bibliography of Regimental Histories* and Neville's *British Military Prints* – unfortunately now out of print. These two volumes will help in identifying most of the ordinary prints likely to be found in the antique markets. The recognition of source is a very important factor, and this can only be acquired by checking the list of regimental histories and approaching public libraries and similar sources to see copies, for many of them have long since been out of print. A third book may be added to the list of indispensables, and that is *Military Drawings and Paintings in the Royal Collection*, for, as remarked above, many of the later prints were copied from original paintings, and Her Majesty the Queen's collection at Windsor is probably one of the best in the world. The above-mentioned book lists and describes some of the well-known plates. A further hazard is, of course, the reproduction of earlier prints in modern histories of regiments. However, since they are almost invariably considerably reduced in scale and the printing processes are fairly easily discernible as being modern, the collector is unlikely to be misled.

Print collecting is, or can be, extremely exciting, for many booksellers and print sellers tend to specialise in one particular subject, and this means that the gaps in their general knowledge may be considerable, which, in turn, means that some of the less-well-known prints may well be offered at very reasonable sums. Persistence is the only hope for the collector and, time permitting, no bundle of old prints or engravings should be neglected. On many occasions unexpected rewards await the ardent collector. Unfortunately many of the prints may be dirty. Tears are repairable, and creases can usually be removed, but the greatest improvement is often obtainable by simple cleaning – the difference can be quite remarkable. Erasers of the very soft type or even semi-stale bread can be used to clean up superficial marks. One of the simplest but most efficient methods of dealing with steel or wood engravings is simply to soak them in lukewarm water. When they are thoroughly soaked they can be gently agitated and then removed and set on a piece of glass and allowed to dry quite naturally and slowly. The improvement can be quite astounding. The cleaning of prints should always be approached with caution, and this warning can be underlined in the case of colour prints. Obviously water-colours are very susceptible, and will run if they become damp.

Some authorities recommend the use of weak hydrogen peroxide to renew the sparkle in colour prints. In most cases this will not harm the print, but beware: at best it may bleach the colour, at worst it may destroy the paper. The treatment applied to these prints may be determined by the condition of the paper. Prints which have long been left in the damp become very brittle and flake, crack, crumble and literally powder away to nothing. In this case there is, unfortunately, no real remedy. Aerosols containing lacquer or varnish can be used to apply a preserving coat, but this does mean, of course, that the print is for ever lacquered!

Mounting the prints is very much a matter of personal choice. Many collectors prefer to leave prints unmounted, and for their storing and preservation transparent plastic envelopes are of great use. These may be obtained separately or in book form. Although they are rather expensive, they are well worth while, since the print is clearly visible and yet, at the same time, will not suffer from constant handling or exposure to the air. If it is desired to glaze the print, then some special adhesive, such as Cow Gum, a rubber solution, should be used, for this will not stain or mark the print should any surplus adhesive come in contact with it.

Page from a typical 18th-century instruction manual showing the exercises of the halberd. Although the halberd ceased to be a weapon of any importance from the latter part of the 16th or early 17th century, a form of halberd, and later a spontoon, was carried by officers as a symbol of rank. This practice was abandoned in the British Army in 1830, though it continued on the Continent.

Plate from *Hungarian and Highland Broadsword*, an instruction book for volunteers published in 1798–99. It was etched by T. Rowlandson.

Captain Hind of the 55th Foot, from a print by John Kay, dated 1790. A known eccentric of the time, Captain Hind walked in a most peculiar fashion.

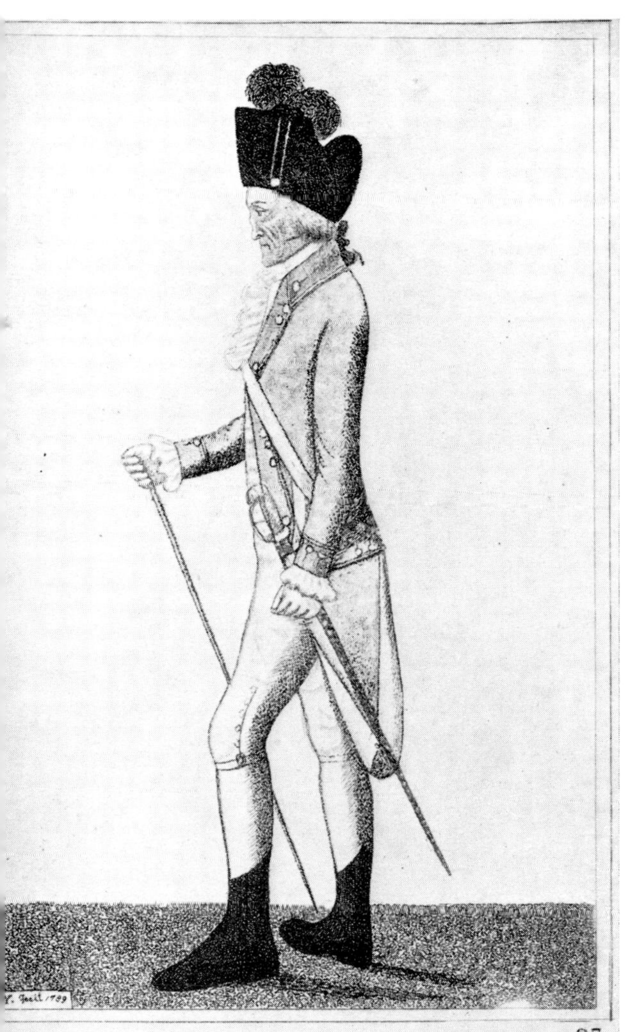

Another Kay print dated 1789, this time of Major Campbell of the 35th Regiment of Foot (Royal Sussex). Many Kay prints will be found coloured, but they were originally simple line engravings.

Print dated 1789 of Colonel Lennox, who was later to become the Duke of Richmond. In 1795 he served as Aide de Camp to George III.

A demonstration of a movement of arms drill, in this case unfixing the bayonet. This is Plate 22 of *The Loyal Volunteers of London*, and shows a member of the Richmond Volunteers, formed in 1794.

The Hon. Francis William Grant of Grant who was at one time Lord Lieutenant of Inverness-shire. In this print by Kay published in 1804, he is reviewing the Inverness-shire Militia and wears a cocked hat.

▼ British Troops during the Peninsular Campaign. Entitled "View on the Tagus Near Villa Belha", this print by William Bradford was published in London in 1809. Such prints were common during the first part of the 19th century and were mostly hand-coloured.

Small line drawing of typical soldier of the latter part of the 18th century with cross belts and shoulder belt plate.

Plate 22

An officer of the King's Dragoon Guards in marching order, from *Spooner's Military and Naval Uniforms*, published 1833–40.

◀

An officer of the Duke of Brunswick's Oels, from No. 16 of *Ackermann's Repository of Arts*, an educational magazine published in 1819. The uniform is dark grey with a light blue upstanding collar, and the shako bears the skull and crossbones badge. The sword is of the general pattern of 1796.

A Cossack officer by John Augustus Atkinson, 1804 – an example of the flood of military art stimulated by the French wars, in which quality varies but, on the whole, accuracy is good.

Two examples of Napoleon's infantry of 1808; coloured plate *c.* 1850. Uniforms of this period were both colourful and impractical, but they did not prevent the French Army from winning a series of amazing victories.

A typical coloured frontispiece from one of Cannon's Regimental Histories published around the 1840s. Depicted is a Sergeant of the 7th Regiment of Dragoon Guards wearing the helmet, with its large bearskin crest, that was popular at this time.

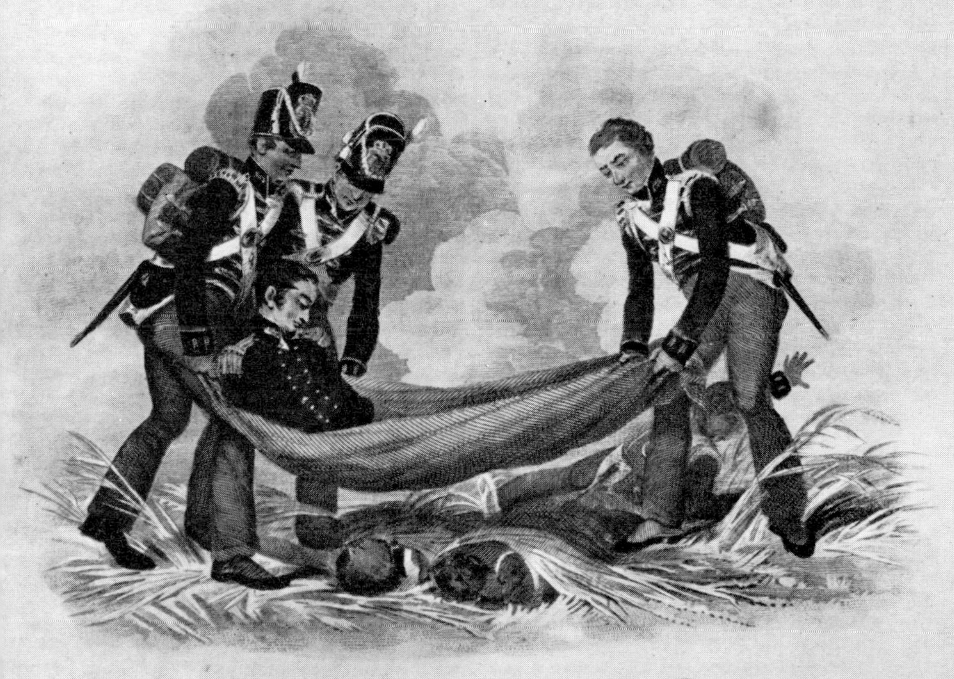

Colonel Gordon mortally wounded, from the *Battle of Waterloo* by T. Kelly, 1816. This accurately engraved plate has been well coloured and is another example of the common practice of removing military plates and selling them as separate items.

C

ıe Second or Royal North British Dragoons, the famous ots Greys, from Spooner's *Costumes of the British ·my*. The paintings were by Michael Angelo Hayes; : engravings by J. H. Lynch. Note that the Scots Greys ›re a bearskin cap rather than a helmet.

British troops storming into Rangoon, another incident ▲ in the Burmese War depicted in the series "The Birman Empire". This is No. 15 of the first series.

ıe storming of the Fort of Syriam, 1824, No. 1 of "The cond Series of Six Coloured Plates Illustrative of the mbined Operations of the British Forces in the rman Empire 1824 and 1825". The paintings were ne from original sketches by Captain Marryat, R.N.

James Grant of Grant, Bart., with a view of his regiment the Strathspey and Grant Fencibles. This is another of several near caricatures by the printer John Kay.

Humorous coloured print of 1804. Such prints were often hire out by print shops for use as decoration at a party or celebration. The military detail is in general correct, but is inaccurate in minor details – for instance, the officer's sword curves the wrong way

The 17th Lancers, the Death or Glory boys, one of the regiments taking part in the famous charge of the Light Brigade during the Crimean War. This is a clear example of R. Simkin's work showing his characteristic style.

Another typical example of the work of R. Simkin done in 1892, showing a group from the Royal Inniskilling Fusiliers in dress and undress uniform. Simkin was meticulous in his search for accuracy, and his work may be generally relied upon as a source of detailed information.

Coloured plate from MacDonald's *History of the Dress of the Royal Regiment of Artillery*, published in 1899. The uniforms were blue; the trousers had a wide red stripe down the side. On the helmets a ball replaces the spike ordinarily used on such helmets.

Coloured plate from Vol. I of *Her Majesty's Army* by Walter Richards, 1890. Two volumes of this work are devoted to the British Army, the third to Indian and Colonial forces. The plates by G. D. Giles are accurate but uninspired.

The Dorsetshire Regiment in full-dress uniform, from a postcard published by Gale and Polden Ltd in the late 19th century. They wear the 1879 pattern helmet; note the chin strap, worn under the chin, as the owner is on duty.

1 2 3

▲

Three commemorative medals of the Metropolitan Police. *Left to Right:* Issued for Queen Victoria's Jubilee, 1897; Coronation medal of Edward VII, 1902; Coronation medal of George V, 1911.

4

◀

Similar medal to that on the top left, issued for the 1887 Jubilee but with a clasp bearing the date 1897, indicating that the holder was on duty for both events.

II

MEDALS

Medal collecting is certainly one of the most popular branches of militaria today. The demand is primarily for good-condition campaign medals, and others, such as those for good conduct, long service or commemoration, orders and similar awards, are less intensively collected. Victoria Crosses are extremely rare, and therefore command a very high price; the 1939–45 Star is still very very plentiful, and fetches only a very low price. Between these two extremes most collectors can find some medals at a price range which will suit their pockets. The value of any medal naturally depends upon a number of factors; rarity and condition play their part, but association is also extremely important. This association may be with a person; an event; a unit; or with other medals which go to make up a particular group.

For the beginner, the collecting of medals can be fraught with hazard, and the only defence is a sound knowledge of military history. Nowhere is this more important than in the case of British campaign medals. The forces engaged on any particular campaign usually comprised infantry, cavalry, artillery, engineers and members of other army departments and, subject to certain conditions, every man who served was entitled to receive the issue medal. In the majority of cases his name and unit were inscribed around the edge, or on the back in the case of stars. Thus one campaign medal may prove to be far more valuable than an apparently similar example simply because it was awarded to a man who served with one small detachment. Fewer medals bearing that unit designation were therefore issued, and their rarity increases the value. The position is further complicated, since a regiment may have been engaged in a campaign, but it may not have been involved in one particular battle. In commemoration of many battles a bar or clasp was added to the medal, so that one unit may have received a bar while another, not engaged, did not, although both units were entitled to the campaign medal. It was not long before some medal dealers and collectors realised the financial rewards to be reaped by removing a commonplace name and unit and substituting a rarer one. This was done by shaving the edge of the medal to remove the original name and then re-engraving a new one, and unfortunately a large number of medals were so treated. The style and correct form of lettering was carefully copied for this renaming, and thus, unless there are any obvious signs of tampering, it becomes difficult to decide whether or not the medal has been altered. However, the obliteration of the original name involved the removal of a measurable amount of metal from the edge. By using a gauge it is possible to detect the small variation in diameter between two measurements taken at right angles to each other. If the gauge reveals that the two diameters are not identical this is a strong indication that the medal has, at some time, been tampered with.

Does this mean that renamed medals should be avoided? The purist would certainly say so. However, if one lacks a particular campaign medal to complete a set there is a great temptation to accept a renamed one as being better than nothing. It should also be pointed out that such renamed examples are worth less than an untouched example, and if a question of any appreciable expense is involved, then very careful consideration should be given to the potential value before making the purchase.

There are in existence nominal rolls giving the names of those who received any particular medal and, at least in theory, it is possible to consult these to ensure that the recipient whose name appears upon the medal did, in fact, receive it. Some societies and libraries possess copies of some of these rolls, but it is not always easy to consult them, for only a few have been widely published.

The condition of a medal is obviously very important, and dealers and auctioneers have evolved a series of classifications to define condition, but it is impossible for every dealer to apply absolutely identical standards, and so interpretation varies one from another. Thus one dealer may describe a medal as "very fine" while another might describe it as "nearly extremely fine", the next highest category. The only reliable method of assessing the accuracy is to compare the description with the item and see how well they correspond. Some medals will be described as "rubbed", and this probably means that at one time it formed part of a group and an adjoining medal has knocked against it and marked the surface. Star-shaped medals are the worst offenders, but a rubbing such as this at least suggests that the medal has been worn at some time.

If one considers the enormous range of medals, naval and military, commemorative and campaign, orders and awards, it may well be difficult to decide what the aim of one's collection should be. Some collectors specialise in a particular campaign, seeking to acquire a complete set representing as many units as possible that took part. Others seek to collect a sequence of medals covering a particular period; others may concentrate on a regiment, while others will merely aim to acquire a representative selection. There is no right or wrong way – it is entirely a matter of personal choice.

Medals are in themselves attractive, and with their ribbons make a pleasant and colourful show, but how to display them is a point of dispute among collectors. One school maintains that they should be left as near as possible in their original condition, even if this means retaining a tattered piece of ribbon. Other collectors prefer to clean them and replace ribbons where required. Again this is a matter of personal selection, but it is reasonable to say that the majority prefer to clean the medal and, where necessary, replace the old ribbon. Certainly they look much better for such attention, and most of the ribbons can be supplied from stock by the larger medal dealers. Excessive cleaning with any form of abrasive will inevitably wear down the surface, destroying detail and impairing the general condition. Soap and water and a very soft brush or cotton-wool may prove to be one of the best and simplest cleaning processes. This is often enough to remove the greater part of surface dirt, and the medal can then be rinsed, and dried. If polish has to be applied it is best to use one of the long-lasting liquid polishes, so that cleaning is kept to an absolute minimum. Silver medals can also be cleaned using one of the proprietary mixtures or a weak solution of ammonia.

The method of exhibiting medals is also a matter for debate. Some collectors prefer to file them away in cabinets sheltered from any possible ill effects of excessive sunlight, others prefer to enjoy them by putting them on open display. Whether open or closed storage is preferred, the medal's condition will be better preserved if it is placed under glass and in a container as airtight as possible. This will greatly reduce the amount of tarnishing and dulling from impurities in the air. Whichever system of display is used, it is as well to secure the medals on some material backing, for this reduces the dangers of scratching and rubbing. If the medals are to be filed away in a drawer, then a plastic envelope makes an efficient container. Full documentation of the medal should always be kept, and it is as well to insure the collection.

One of the joys of medal collection is the possibility of discovering the "sleeper" – the unrecognised bargain. It is becoming increasingly less common to do so, but there is still just sufficient chance for it to be a worthwhile hope. The collector needs to persevere and dig hopefully into every pile of bits and pieces, every odds-and-ends box, as well as looking over all the junk stalls in the market. Auction sales devoted to coins and medals are fairly frequent these days, and many collectors prefer to make their selection from the auction room or a dealer's stock. There is no doubt that a dealer of repute is a guarantee of quality, but, of course, the price demanded for any medal will reflect this standard.

The history of medals is a very long one, although the idea of a general issue is comparatively modern in origin. Certainly in Britain it was not usual to make a general issue to all men and officers before the middle of the 19th century. There are suggestions that commemorative medals date back to the 16th century at least, and there are in existence some such medals which are usually described as Armada Medals. It does not seem at all likely that these were generally issued but were reserved for a few of the commanders, and they were probably in the nature of a personal award from the sovereign. This particular medal is of gold or silver and is a large oval shape; on one side is the Queen, on the other side an ark in the flood. Queen Elizabeth is also credited with having issued at least two other medals, but it is not at all clear as to the basis on which these were issued. It seems fairly certain that they were worn suspended from a ribbon around the neck in much the same way as some present-day orders.

In 1643 Charles I issued a Royal Warrant which proclaimed the issue of medals, under strict conditions, to members of the *Forlorn Hope* who led the attack. These were of silver gilt, but it was not until 1650, during the Commonwealth, that Parliament voted to issue a medal to all those who took part in the great victory at Dunbar, and both officers and men were to receive this. There were subsequent periodic issues of medals during the 17th century.

Throughout the 18th century a number of what might be called private medals were issued by various individuals or communities, but none was a general issue. In 1816 the Prince Regent commanded that every officer and man who served at Waterloo should be issued with a medal. He also commanded that "the ribband issued with the medal shall never be worn but with the medal suspended to it". It was the head of the Prince Regent that decorated the obverse, not that of George III.

In June 1847 a belated general order authorised the issue of a Military General Service medal covering service during the period from 1793 to 1814. Since the medal was issued to cover such a long period, there were naturally many battles and campaigns to be commemorated, and in all nearly thirty bars were issued. Few, if any, soldiers would have qualified for all the bars, and the highest number received by any man seems to have been fifteen. A similar medal – The Naval General Service Medal, authorised in 1847 – was to cover actions at sea during the Napoleonic Wars, and for this there were 231 different bars. Having set the pattern, the issuing of campaign medals has continued in like fashion to the present.

British campaign medals are probably the most plentiful, because the British Army was engaged in so many campaigns throughout the 19th century. In the case of the United States there are far fewer campaign medals, excluding those of the Second World War, only some twenty-six covering campaigns as varied as the American Civil War 1861–65 and the Haitian campaign of 1919–20. There were, of course, a number of U.S. awards for bravery, such as the Distinguished Service Cross, the Distinguished Service Medal and the Distinguished Flying Cross, although most of these were initiated only in the First World War. The Second World War saw a great increase in the number of campaign medals issued by all the combatants, and these are still very plentiful and easy to acquire.

Decorations and awards form a special group and need not be specifically military, and therefore some fall outside the scope of this book. However, awards for acts of military bravery are obviously rarer than campaign medals, simply because fewer were issued; for the same reason officer's medals are scarcer than those of other ranks. The Victoria Cross was instituted during the Crimea, and this medal has become increasingly more difficult to acquire than it was originally. In the Crimea and Indian Mutiny considerably more crosses were awarded than during later campaigns, when an equivalent deed merited a lesser award.

In addition to medals awarded for bravery there were many issued for long service, as shooting prizes and romantic ones such as those presented to North American Indians by the British Sovereigns from George I to Victoria, but these later ones are very rare.

Even after this brief survey of medal collecting it must be very apparent that knowledge is all-important. Valid reference books are essential, and for general information on British medals it is difficult to better Gordon's *British Battles and Medals*; while for a general survey Dorling's *Ribbons and Medals* will prove indispensable. Despite all the potential pitfalls, medal collecting is a very satisfying and stimulating aspect of militaria.

Campaign Medals with bars of the first Sikh War, Sutlej Campaign, 1845–46. "Army of the Sutlej" is engraved on the edge of the medals.

BRITISH AND COMMONWEALTH MEDALS: (1) Victoria Cross; (2) Distinguished Service Order; (3) Distinguished Service Cross; (4) George Medal; (5) Army of India Medal; (6) Waterloo Medal (1815); (7) New Zealand Medal (1845–66); (8) South African Medal (1877–79).

MERICAN MEDALS: (1) Distinguished Flying Cross; (2) Bronze Star;
) Air Force Cross; (4) Air Medal; (5) Joint Service Commendation Medal;
) Purple Heart.

FOREIGN MEDALS: (1) W.W. I War Cross (It.); (2) Military Medal (Fr.); (3) 1939–45 Comm. War Medal (Czech.); (4) Order of the Red Banner (U.S.S.R.); (5) The 1916; Black & Tan Medal (Eire); (6) Cross of War with Palm (Bel.); (7) 1939–45 Iron Cross 1st Cl. (Ger.); (8) Order of Vasa (Swe.).

OREIGN MEDALS: (1) Order of Leopold (Belgium); (2) Order of the olden Fleece (Austria); (3) Order of Merit (Chile); (4) Order of the rown (Italy); (5) Legion of Honour (France).

1

2

3

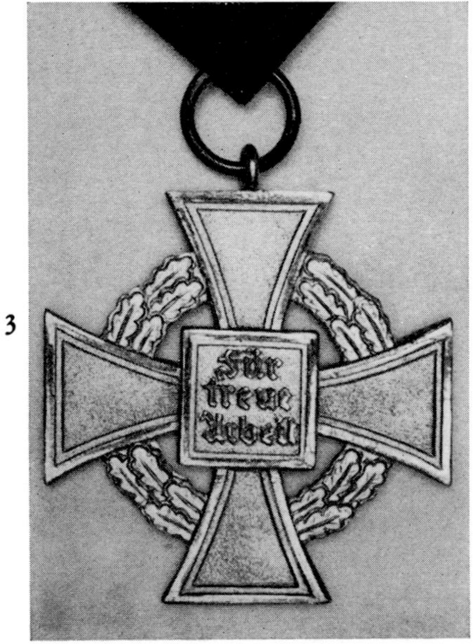

(1) Iron Cross, first issued during the First World War and reintroduced by Hitler 1st September 1939.

(2) Africa Star, Second World War, issued to service men and women taking part in the African campaign between June 1940 and May 1943.

(3) The reverse of the Nazi 50-year service award medal (see next page), bearing the words *Für Treue Arbeit* – "For True Labour".

(1) Solid silver Knight's Cross of the Nazi War Merit Crosses, awarded only on the authority of Hitler, who also presented it.

(2) Nazi medal awarded for services to the Party. This example bears on the back the words *Für Treue Dienste* – "For True Service". (3) Nazi cross awarded for long and faithful service; a liberal award that was extended to employees of private concerns.

2

3

Award for officers and men of the German Air Force Anti-Aircraft Artillery, instituted by Herman Goering January 1941.

Spanish Volunteers in Russia Medal, issued to a number of Spanish volunteers who served in the Blue Division during the Russian Campaign, 1941–42.

Three War Service Crosses, the civilian counterpart of the Iron Cross, though it was awarded to the military and civilians alike.

The Guerilla warfare badge, instituted in 1944 and awarded to members of the Armed Forces of Germany who were fighting against partisans in the Balkans.

Royal Artillery Kettle Drummer's Coat, *c.* 1750.

III

UNIFORMS

A sound knowledge of uniforms is one of the basic essentials of almost every collector interested in militaria. For the miniature-soldier enthusiast seeking to produce accurately painted models, for the collector of uniforms – indeed for anybody interested in any item of military equipment this knowledge of uniforms is a first requisite.

For details of uniform prior to the mid-18th century most of the information can be culled only from portraits, paintings, prints and written descriptions. Even with all these primary sources, it is not always easy to be certain that one has built up an accurate impression, for so much depends upon the accuracy of the artist. With verbal descriptions the same problem exists, probably to an even greater degree, for it is generally easier to paint a uniform than to describe it.

In Britain the position becomes a little less confused from the middle of the 18th century, and with the advent of the 19th century matters become much clearer. The British War Office was by then producing sets of orders defining, in detail, exactly the uniform to be worn by any one regiment or group of officers. For the collector primarily interested in 19th century uniforms these are the most valuable source books, but unfortunately the Dress Regulations, as they were called, were issued at irregular intervals and are now scarce. Such warrants and regulations had been issued at earlier periods, but few have survived. Pictorial evidence is also far more readily available for this period.

The origins of heraldry during the latter part of the 12th and early 13th centuries led to something approaching a style of uniform, for many of the nobles encouraged their followers to bear their arms or part of the arms upon the shield or tunic. This practice was fairly general throughout most of the Middle Ages, but in Britain was stopped by Henry VII, who, determined to control the small baronial armies, rigorously implemented the Statute of Liveries. He was quite capable of imposing very heavy fines on any noble who transgressed this statute.

In times of national crisis the British ruler was entitled to call upon his subjects to muster and report for duty, and certain lords and towns were responsible for supplying a number of armed soldiers, and such contingents were often clothed in a particular style of costume. However, it is during the 16th century that the origin of uniform as we understand it today may first be discerned.

Although red is thought of as the traditional colour of the British soldier's uniform, it did not, in fact, become common until the end of the 16th century. Throughout the whole of the 16th century and even during the latter part of the 15th century red crosses on a white background were commonly adopted by British troops serving both at home and abroad. At the end of the 15th century the Yeoman of the Guard are described as wearing white and green jackets, and white remained a very popular colour during much of the 16th century, although one would have thought it very impractical for war wear. Throughout the rest of the 16th century there are numerous mentions of men clothed in red, blue, green and gold; even yellow is recorded for a contingent from Canterbury, which seems to have been a little unusual. It must also be made clear that these standard colours were used only for the jacket, while caps and hose were still freely chosen by the individual, and there are references to hose with one red and one white leg, each with a stripe down the side. Caps were red or blue, and some were decorated with large plumes.

By the early 17th century the system of distinguishing regi-

Other ranks coat of the First Foot Guards, *c.* 1790.

ments by means of favours was well established. These bunches of coloured ribbons were worn at the shoulder, the knee or the boot. Distinctions were also made by a sash or scarf, and many a soldier owed his escape after a disastrous battle to the ease with which a sash might be removed, leaving him anonymous among the victorious army. During the British civil wars of the early 17th century, troops on both sides were frequently dressed in very similar styles, and the main distinction was by ribbons around the hat. Cavaliers tended to favour red, while Parliament favoured orange. It is during the same period that the custom of wearing regimental coloured coats became well established. The majority of the King's men seem to have favoured red or blue, but there are many references to yellow, grey, white, purple and green. These coats were lined with a different coloured material, which became visible when the cuffs were turned back, and later, when lapels were introduced, these also clearly showed the colour of the lining. These different coloured edgings to cuffs and collar were known as "facings" and became an important distinguishing feature of later uniforms. Since both sides wore such very similar uniforms, commanders often felt it necessary to add some further distinction, and these field marks were usually fashioned from things readily to hand, such as green leaves, a white handkerchief or a piece of paper pinned to the hat or sash.

It is with the return of Charles II that one may formally date the beginning of the British Army. Already the uniform was beginning to take on its later form as the short jacket previously worn by musketeers was gradually lengthened, and by the beginning of the 18th century it had extended to the knee. At first these coats were simple and loose fitting, but towards the end of the century they were being shaped and given a waist. Most of them seem to have been decorated with a profusion of buttons, and for officers at least, the button-holes and surrounding area were embroidered with gold and silver or occasionally fitted with button loops.

It would appear that the Duke of Marlborough was one of the pioneers seeking to standardise and reorganise the very chaotic conditions applying to uniform and equipment in the British Army. He certainly introduced the system of sealed patterns, whereby a style was approved and a sample of the material was examined, sealed and used as a standard by all contractors supplying that uniform. In 1707 a Board of General Officers was set up, and one of their tasks was to see that uniforms made on contract were indeed up to standard. They also laid down certain regulations as to the equipment that should be issued to the foot soldier. Essentially the cavalry uniform was the same as that for the infantry, but their coats usually had very wide deep skirts. The cuffs were decorated with facings and occasionally edged with lace, while around the neck was worn a white neck cloth or cravat. Gold and silver fringes were commonly added to sashes, waistcoats and gloves. Distinctions of rank were shown by the amount of lace or ribbon affixed to the jacket.

By the early 18th century the majority of British soldiers were dressed in a uniform which comprised a pair of thick leather shoes, gaiters over white stockings, breeches to the knee – the colour of these depended upon the regiment – a waistcoat and over this a red coat lined with another colour material. Cuffs were very wide and full, frequently almost to the point of absurdity. The long skirts of the overcoat were often fastened back to give the familiar looped appearance.

One important change which took place in the British Army around the middle of the 18th century was the recognition of the practice whereby regiments came to be known by a number rather than by the Colonel's name or any other designation. They were all placed in order of ranking and given a number, which then appeared on hats and other items of equipment. This practice seems to have been used by many regiments prior to this period, but on a very casual basis.

Towards the end of the 18th century the soldier's top-coat

underwent several sweeping changes. No longer did it droop open gracefully, secured only at the neck, but was now buttoned to the waist and fitted with a collar that stood up, and the skirts were considerably shortened. Waistcoats, which had previously nearly all been red, were also replaced more and more by white ones. Trousers were now most often white, and as a whole tended to be far tighter-fitting than in previous years.

Maurice, Count de Saxe writing, in his *Memoirs Concerning the Art of War*, translated from the French in 1759, said, "Our Dress is not only very expensive but most inconvenient, the soldier is neither shod nor clad." Certainly the greater part of the 18th-century uniform seemed to have been designed largely for show, with hardly any thought as to practicality. It is ironic that more than 130 years afterwards General Viscount Wolseley, writing in a book entitled *The Armies of Today* (1893), should say, "It is however curious to note, that for the hard marching and many bodily exercises which fall to the soldier's lot on active service our army was more suitably dressed in the reign of William III than it has been generally this century."

The Napoleonic Wars soon forced a number of radical changes in the design of the British soldier's uniform which were, on the whole, towards a simpler, more utilitarian design. One of the changes was the abandonment of the long gaiter with anything up to a dozen or sixteen buttons at the side, and in its place was used a small one, rather like an ankle gaiter. Soon even this was abandoned and ordinary trousers substituted. The soldier's jacket tended to lose its long tails, and was now cut short at the front and buttoned to the waist. It was also at this time that green made its appearance among the rifle regiments.

Once the wars were over, this tendency to simplification was somewhat reduced, and laces, ribbons, double fronts and a variety of other gaudy accessories began to reappear. Foreign uniforms were copied and ideas adopted, and in the case of Lancer Regiments the entire uniform was copied from the Poles. However, it was with the coming to the throne of the Prince Regent as George IV that uniform mania became rampant. Fashion-conscious as he was, George IV displayed a keen interest in military uniforms. His ideal seems to have been a tight-fitting uniform, and so tight did some of the jackets and trousers become that the troopers must have found it extremely difficult to move their arms or legs at all without disastrous results. Possibly the most elaborate of all were those of the Hussars regiments, with their great tall fur busbies fitted with plumes and bags, elaborate gold-braid plastron fronts and a pellise or extra jacket dangling from their left shoulder. Tight white trousers completed their picturesque, but rather impractical, uniform. It had now become standard practice to have two types of uniform, the dress and the undress. Dress uniforms were reserved for levees, dress reviews, Drawing Rooms and Birthdays; undress, for general wear.

It was William IV who set down the guiding principle that the entire British Army was to be clothed in red, with the exception of Artillerymen and Riflemen. Regiments of the line were ordered to wear gold lace, while the militia had silver. Experience during the long reign of Victoria, with its numerous minor frontier wars as well as the Crimean, gradually forced upon a somewhat reluctant or reactionary high command the necessity for simplification of uniforms. The American Civil War, too, had clearly illustrated the impracticality of colourful and elaborate uniforms upon the modern battlefield. There were experiments as early as 1840 with dull-coloured uniforms, and during the Abyssinian campaign of 1868 a drab, dusty, khaki colour had been used for some of the uniforms. Many of the Militia and Volunteer Regiments raised around the middle of the 19th century also abandoned the elaborate coloured uniform and tended towards dull greys and dark blues, but it was probably the South African War – The Boer War – which drove home once and for all this vitally important lesson. In 1902 the army introduced a general service dress for wear on most occasions, but retained the full-dress one. The First World War saw the standard adoption of khaki as the traditional colour for the British uniform from that time forward. From this time on the story of uniforms continues, in essence, to be that of a change towards more practical design not only in the British Army but in all areas.

The Second World War saw the introduction of the smock, the zip and a general easiness and casual approach to uniform. There was also, following upon the Second World War, possibly as a reaction to the utility uniform, a reintroduction of more colourful dress or walking-out uniforms, and at present there is something approaching a reasonable compromise. Traditional uniforms were retained for many ceremonial purposes – in Britain the Household Cavalry; in Denmark the Royal Bodyguard; in America at West Point; in France St Cyr and the Guarde Républicain – are all instances of this lingering tradition.

The history of the uniform of the Royal Navy is much shorter than that of the Army, for it was not until 1748 that any official uniform for officers was adopted. Essentially the uniform was an equivalent to the civilian dress, and variations have, generally speaking, been less than for the Army. Blue has remained the basic colour from the beginning, and rank was usually distinguished by the arrangements of buttons. Until the early part of the 19th century there was little distinction between the appearance of naval uniform and that of civilian wear. A long top-coat with wide skirts, folded cuffs and lay-down collar was the general basic style. For full dress there was naturally a great deal of gold embroidery and lace, but during the first part of the 19th century there was a general simplifying, with the front of the coat being buttoned to the waist and with a general reduction in the size of the skirts. By the middle of the 19th century the basic uniform consisted of a frock coat with sundry epaulettes and loose-fitting trousers.

Broadly speaking, the collector of uniforms cannot reasonably hope to acquire anything pre-dating the 19th century. Certainly earlier pieces do turn up, but they are invariably expensive, and in such cases a second opinion can reasonably be sought. Many of the uniforms in auction rooms date from the mid-19th century usually of militia or yeomanry regiments.

For the collector of uniforms one of the greater hazards lies with the products of the theatrical costumier. Film wardrobe managers and those of the large theatres feel that no expense is too great to acquire authenticity, and many of their uniforms are, in fact, facsimiles. It so happens that occasionally these pieces find their way on to the general market, and the collector may often be sore put to the test to decide whether a piece is genuine or not, especially when the piece is dirty or shows signs of wear. To decide whether it is genuine or not is no easy matter. The invention of the sewing machine dates from around the middle of the 19th century – there were earlier examples dating back to 1790, but these were largely unsuccessful – and the generally successful ones date from the 1840s and 1850s. Obviously a uniform pre-dating this period would have been hand stitched. The regularity and evenness of the stitch will give some indication as to whether it is hand or machine done. Material, too, will give some indication, and quite obviously the uniform must be correct in detail – to know which one must have access to reliable textbooks. Many of the pieces will bear labels or at best show signs of one having been there. Many of the reference books give some information as to the known suppliers of military uniforms, and again this may give a little guidance. Less easy, but certainly as important, is comparison with known genuine examples, for many Regimental and National museums listed in the appendix (see pages 233 to 249) contain a number of genuine and authenticated uniforms. It is not always easy to examine these closely, but even from a glass case one can take pointers as to how the lace is attached, how button-holes are made, the type of hook and eye – all points which can help in arriving at a decision.

Over the past fifteen years there has been a growing interest in

Nazi uniforms. At this distance in time, some twenty years after the end of the war and considering the size of the forces involved, surprisingly little Nazi material has survived. Much was deliberately destroyed after the war, and the strict rules which pertained to the displaying of Nazi emblems and uniforms also took their toll. However, some items still turn up, usually caps and tunics, and they are still, at the moment, reasonably priced. This demand has, as always, stimulated an increase in the output of books dealing with the German armed forces, and improving knowledge will naturally further stimulate demand. Value of such pieces is normally dictated by condition, but in general most of the German stuff is good, and therefore the prime consideration is of rarity. Reference to the appropriate source books will give some guidance on this matter.

Cleaning of uniforms is very much a matter for experts, and with earlier pieces it is strongly recommended that advice is sought before any action is taken. Modern methods of cleaning and restoration are extremely efficient when correctly applied, but care is essential.

Display is another problem. Ideally a dummy makes the best show piece of all, but for the modern householder or flat dweller the problem of space precludes many of these. Possibly the best compromise is the well-shaped clothes hanger and the transparent plastic bag. This enables the tunic to be seen, but, at the same time, preserves it from the ravages of moth, dust and air pollution. This problem of storage and display has meant that although there are many well-known collectors who possess hundreds of fine examples, uniform collecting has never been one of the most popular branches of militaria.

Lt. General's Coatee – Army Staff, *c.* 1813.
Rank is indicated by a crown, and the gilt buttons carried a crossed sword and baton encircled with laurels.

Captain's full-dress tunic of the Cumberland Regiment, *c.* 1856. Rank is designated by the crown and star on the collar, although this was later changed to two stars. The tunic has yellow facings, silver lace and buttons that bear the number 34 encircled in laurel leaves.

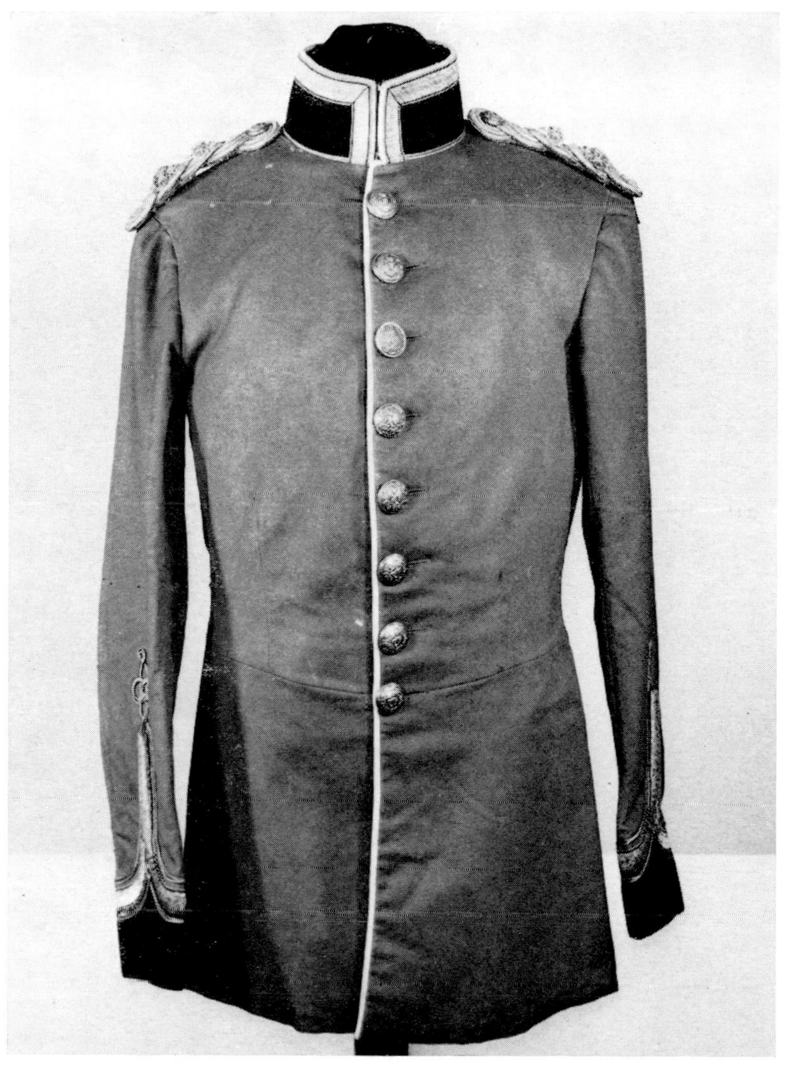

Officer's full-dress tunic of the Royal Dublin Fusiliers, *c.* 1901. This regiment had frequent changes in the colour of its facings, but at this period those of the 1st Battalion were scarlet, while those of the 2nd Battalion were faced in blue.

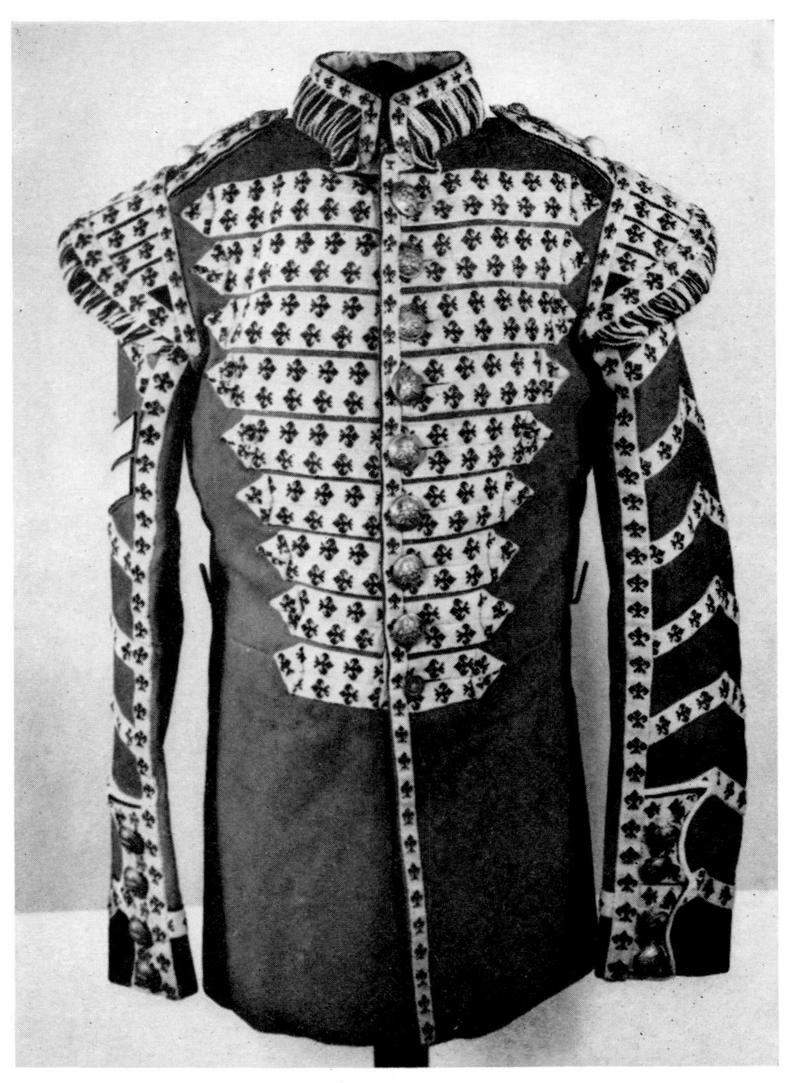

Full-dress tunic of a Grenadier Guards Band Corporal, *c.* 1913.

Khaki tunic of the Captain of the Connaught Rangers, First World War. The badges of rank, originally worn on the sleeves, were later transferred to the epaulettes.

QUEENS WEST RIFLE VOL
QUEENS WESTMINSTER RIFLE VOLRS

Epaulettes of plain Sheffield plate metal worn by a Kent volunteer group during the Napoleonic wars.

◀
Cartridge box and belt worn by a private of the Queen's Westminster Rifle Volunteers; late 19th century. The portcullis on the badge is from the Arms of Westminster.

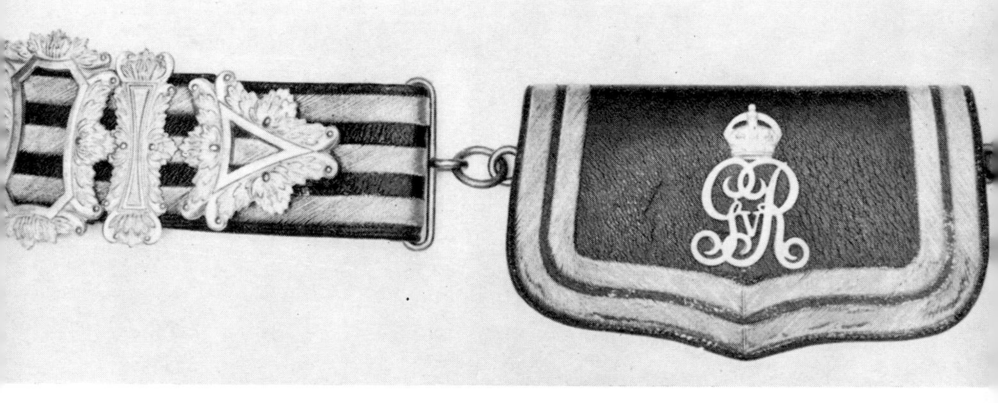

Pouch and belt of Royal Medical Corps, George V, 1910–36. The belt is composed of alternate black and gilt stripes, and the Royal Cypher is done in gilt on a black leather pouch.

Waist belt of the Army Veterinary Department. The cypher and oak leaves are of silver, mounted upon a frosted gilt buckle.

Cross belt with pouch bearing the Royal Cypher and a crown – the Army Pay Department; dating from the latter part of the 19th century.

Top: Belt buckle of German Infantry – First World War. *Lower Left:* Belt buckle of Police of North Rhine and Westphalia Province. *Lower Right:* Belt buckle of German Infantry – Second World War.

A 19th-century sabretache, probably of Turkish or Eastern European origin. These large pouches were intended to hold messages or serve as writing surfaces. Shown is the interior, which is functional to the extent that it contains a number of pockets. (See facing page.)

Shoulder pouch worn by an officer in the Indian Army. Such officers indulged in personal adornment that would have been unacceptable to British Army officers. In place of the conventional leather this pouch is of brass covered with intricate silver inlay pattern work.

Exterior view of sabretache pictured on the facing page. As time passed such articles became more elaborate and less functional. The quality of decoration here is of a very high standard, done in silver thread on a blue background.

POSITIONS IN THE MANUAL EXERCISE,

AS TAUGHT AT HYTHE.

8. TRAIL ARMS.

10. BAYONET EXERCISE AGAINST THE SWORD.

Sword movements "as taught at Hythe", the Small Arms School in Kent. Typical plate of engravings that were common in the volunteers' handbooks of the mid-19th century.

9. STAND AT EASE.

11. BAYONET EXERCISE AGAINST THE SWORD.

IV

BOOKS AND EPHEMERA

For the collector of militaria good reference books are essential, but many military books have an intrinsic value and interest of their own. Leaving aside the purely reference books, most others fall fairly clearly into one of the following categories: Instruction Books, Regimental Histories, General Histories, Accounts of Campaigns and Official Publications.

Instruction Books have been briefly discussed in the section dealing with prints and, as was pointed out there, some were published as early as the 16th century. Early examples are obviously going to be expensive, but those dating from the mid-19th century, even some of the 18th century, may still be found at quite moderate prices. Some are, for the non-specialist, rather dull, being concerned only with the movements of infantry and cavalry, and the plates are covered with little rectangles representing groups of soldiers and the way they move.

Regimental Histories constitute one of the largest groups of books. Most collectors tend to concentrate on one or two particular regiments or armies, but a few aim to build up sets of regimental histories. As a general rule, it can be safely reckoned that the earlier the regimental history, the more expensive it is likely to be today. As pointed out in the chapter on prints, a very large number of these regimental histories were produced by R. Cannon during the first half of the 19th century, and these together with their coloured plates are much sought after.

Some of the more modern histories have been sponsored or subsidised by Regimental or similar funds, and are therefore sold at a very reasonable figure indeed. As a rather sweeping generalisation, it is probably true to say that the majority of newer publications are more concerned with recent rather than past history, but there are some exceptions, and these cover the history of the regiment from its inception to the present. Recent amalgamations and re-organisations which have taken place in the British Army have caused the disappearance of many old and long-established regiments. These changes have stimulated particular interest in these units.

In the field of *General Histories*, Sir J. W. Fortescue's *History of the British Army* still stands today as one of the most authoritative accounts, although even this great work is not without its failings and minor inaccuracies. Other general histories have been published, but they are inevitably briefer and often rely heavily upon Fortescue. Many of these general histories are listed in the bibliography, and most collectors will be familiar with them.

Campaign Accounts are a somewhat specialised taste, but if a collector has a special interest in one regiment, then it is always worth checking histories of battles in which it took part. There may well be gems of information concealed within the covers of the dullest-looking tome. Often there will be in such an account a mention of some specific soldier, officer or article of equipment that may well provide a useful lead in sorting out a doubtful point of fact. Again photographs and prints are often included, and these also may provide incidental information.

The broad heading of *Official Publications* can be taken to include such diverse items as Black Type copies of Acts of Parliament dealing with the raising of the Standing Army of the Commonwealth or instructions for the armourer issued during the Second World War. As far as the British War Office is concerned, they have, over the centuries, issued a number of publications dealing with a variety of topics ranging from equipment, dress, general regulations, barracks, weapons, to military law and finance. For the collector of militaria and the model-soldier

Poster issued in 1803 calling upon the men of an area in the county of Somerset to form a troop of cavalry and another of infantry to resist the imminent French invasion.

Defence of the Country.

TO the Gentlemen, Yeomanry and others of the Parishes of *Wiveliscombe*, *Huish-Champflower*, *Fitzhead*, *Lydeard Saint Lawrance*, *Stogumber*, *Tolland*, *Monksilver*, *Nettlecombe*, *Saint Decumans*, *Old Cleeve*, *Elworthy*, *Brompton Ralph*, *Clatworthy*, *Chipstable*, *Raddington*, and all adjoining Parishes.

AT a period so interesting as the present when our most enveterate and implacable FOE is expected every Day to invade us, it becomes the sacred Duty of all Loyal Subjects to come forward in Defence of our beloved King and most glorious Constitution.

THE Energy and Patriotism already displayed in the County of SOMERSET, embolden me with the Hope of commanding a TROOP OF CAVALRY.---I therefore take the liberty of requesting that the Gentlemen, Yeomanry and others, (not already engaged in a Military Capacity,) will enroll themselves under my Command for the preservation of our Families, our Property and that Liberty so justly dear to every true-born Briton.

T. GORDON.

N. B. BOOKS Open for Enrollment at MIDDLETON COURT, and at Mr. HANCOCK'S, of FORD, near WIVELISCOMBE.

Middleton Court, August 1st, 1803

CONSIDERING it an indispensible Duty which every Man owes to his King, to his Country and to himself at this momentous Crisis, to shew his readiness in assisting to counteract the wicked Designs of our rapacious Foe, and not doubting the Loyalty, Good Sense or Zeal, in so laudable a Cause, of the Inhabitants in the Parishes adjoining to *Wiveliscombe*, I will except Inrollments of such of the Inhabitants of the above mentioned Parishes as are willing to serve in the INFANTRY with the Inhabitants of *Wiveliscombe* who have honoured me with their Choice as an Officer.

LACY YEA.

Oakhampton House, August 1st, 1803.

N. B. ALL those in the Parish of *Wiveliscombe* who have agreed to serve in a Military or other Capacity for the Defence of the Nation, or Removal of Property, or who are now willing t[illegible]nroll themselves are desired to meet in the *GREEN*, on *Wednesday* the 3d of this present *August*, by 5 o'Clock in the Afternoon.——The Inhabitants of *Huish Champflower*, *Clatworthy*, *Chipstable* and *Raddington* are desired to meet in the CHURCH YARD of *Huish Champflower*, on *Thursday* the *4th* of *August*, by *six* o'Clock in the Afternoon, and the Inhabitants of *Brompton Ralph*, *Tolland* and *Fitzhead*, at PITSFORD HILL by *six* o'Clock in the Evening of *Friday* the *5th* of *August* Instant, to be inrolled either as CAVALRY or INFANTRY.

PILES, PRINTERS, NORTON.

enthusiast, probably the most useful are the dress regulations, but these are not common, and are eagerly sought after, particularly those of the late 19th century and early 20th century. During the early part of this century the War Office, or publishers with its official approval, issued a large number of small pocket booklets, some four inches square, dealing with a whole range of topics. These included instructions to armourers on how to repair and check weapons, manuals of artillery practice, manuals of first aid, pay regulations – in fact, they covered almost every aspect of military life. These are still readily available at fairly low prices, and are always worth adding to one's military library. For the collector of such things as military bayonets the *Instructions to Armourers* are particularly useful, because many of them have full tables showing the markings on bayonets issued to a very large range of units and depots.

Perhaps not strictly official publications, are such items as those illustrated on pages 83, 89 – proclamations, notices, returns, souvenir programmes, etc. Since these items are rather specialised, they are frequently outside the experience of many print dealers, and the lucky collector may well acquire them at a very moderate figure. Army Returns such as that illustrated on page 93 are less frequently encountered, because they are official documents, and were therefore normally retained at headquarters, but a number of them, and such things as commissions, do find their way out on to the general market.

One small but important point that is worth bearing in mind whenever looking at any form of military book or document – check the fly leaves for signatures or names. Many books bear inscriptions as on page 90, and it is always worth checking the identity of the owner, for this may well add an association value far in excess of an ordinary unsigned copy.

There is a very encouraging indication that publishers are showing an interest in reprinting some of these early books, and it is to be hoped that this practice will continue, so enabling military enthusiasts to build up a library of source material at a reasonable figure.

Another type of printed military item that is a useful source of information is that of the military magazine. Many of these were issued by commercial publishers for sale to the public, while others were internal publications of a regiment or similar group. Some of them were no more than social news sheets, but many of them carried photographs, drawings, comments and quite interesting articles. Some, like the *Cavalry Journal*, were published for many years, others had a much briefer existence, but all are worth checking and looking at. During both World Wars certain areas published their own newspapers, and again these are rapidly acquiring the status of collectors' items. There are also the learned transactions of such bodies as The Royal United Services Institution.

Still within the realms of printed material are the postcards. Quite apart from the series illustrated on page 41, which were very popular during the late 19th and early 20th centuries, there are many other souvenir cards illustrating types of national soldiers, scenes of battles and so on. There are also some issued by museums, such as those from the Edinburgh Museum of the Pilkington-Drummond figures which make a very fine sequence illustrating the history of British military uniforms. Popular, especially in France, are postcard-size productions which carry a repetitive sequence of foot or cavalry soldiers, usually in rather crude but not unattractive colours. Smaller in size, but possibly of greater interest for the militaria collector, are the several sets of cigarette cards issued at the beginning of the century. Most of them are extremely attractive in their own right and carry very full and accurate information on military headdresses or military uniforms. John Player & Sons Ltd did a set of fifty on military headdresses which are most attractive and very accurate. The same firm, Players, also did a series of fifty on Riders of the World, and some cavalry units are included among these. Carreras Ltd also did a very good series of fifty dealing with army uniforms, with small, accurate and quite pleasant illustrations limited to the history of the British Army. Other series dealt with Naval topics and ships, and reference to appropriate books will supply other titles.

Turning to more original items, there are a large number of objects that can, at best, be described as military curios. Many such objects have little intrinsic value but rather association value, and this raises some very large queries. The criterion of any item with association value must lie in the proof of that association. To say that such and such a sword belonged to one particular officer does not prove that this is so! It may well have done, but being able to prove this statement is a very different matter. Similarly, one small cannon ball looks very much like another, but whether it was found at Waterloo or Inkerman or Gettysburg or Coldharbour is a very difficult matter to prove. Collectors should be wary of any such association items unless there is a substantial means of proof available. Then, too, there are the items that are not associated in the sense that they have a connection with one specific person, but are rather general items associated with an event. At Christmas 1914 Princess Mary sent gift boxes to troops serving at the front, and these contained pipes, tobacco, cigarettes, Christmas cards, photographs and so on. These were frequently kept by the troops as souvenirs, and may still be found complete and unopened. Other forms of souvenirs are not at all uncommon, and they seem to have become more popular from the Boer War onwards, although examples dating back to the Crimean War or earlier may be found. They are frequently made from cartridge cases which are fashioned into cigarette lighters, ashtrays, miniature caps or even such unlikely objects as frames and letter racks. They are not popular with collectors, and are generally of little value. Far more rare and expensive are the early examples of what were probably originally souvenirs, for example, the carved powder horns. These powder horns of the American Revolutionary War often have a map carved on them and some inscription. They are rare, and authentic ones command very high prices; so high in fact that doubtless the forger has been stimulated to produce one or two.

Another interesting group of objects are those described as prisoner-of-war work. During the Napoleonic Wars large numbers of French prisoners were detained at centres throughout Britain; Liverpool, Edinburgh, Dartmoor, Portsmouth all had their quota, as did many other places. While the prisoners were not harshly treated, they were certainly not pampered, and in order to earn some money they were allowed to make and sell items to the local populace. All of their work was done using the simplest of material – bone, bits of leather, their own hair, straw and odd pieces of wood. Some were skilled craftsmen, and one of their finest productions was the full-rigged ship – these are beautifully made, probably by sailors who knew their subject, and are often rigged with the maker's own hair. It is unusual to find these in good condition, and when they do turn up they command a very high price. Similarly, small automata in bone and wood were produced by these prisoners, and again these are naturally expensive. Much more likely to fall within the average collector's range are smaller items, such as sets of dominoes in rather crude wooden boxes, each domino carved from a piece of bone, small toys, crucifixes, and rings or watch-chains. Many of the items ascribed to prisoners of war are often questionable. Some Tonbridge work is so described, but, in fact, Tonbridge ware was in its heyday some thirty years after the Napoleonic Wars.

There are still many other rather off-beat fields that have not been mentioned, since they are essentially specialist tastes with a military connection. War stamps, wartime money, propaganda leaflets are just a few of these.

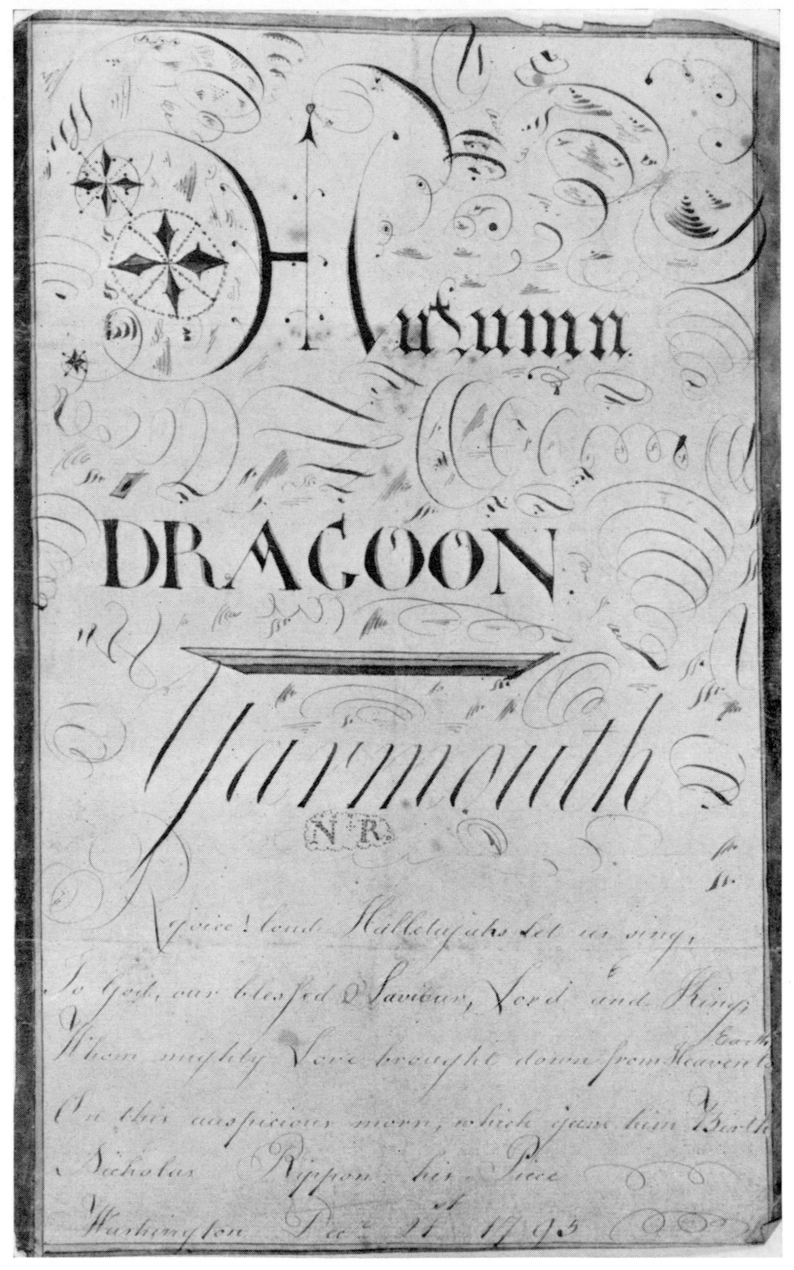

Specimen page of handwriting styles. The word Dragoon suggests a military origin, but it is simply an example of one style.

This Precept was issued to the Chief Constable of the Division of Langbaurgh West on the [illegible] Day of January 1831.

Henry Hirst
Clerk of the Gent. Meetings at [illegible]

No. 1. Recd. from the Constable Office in Stokesley Saturday, Jany 15th 1831

Geo Brigham

MILITIA.

Form of Precept to the Chief Constables of Hundreds, or other Divisions, for Issuing Orders to Petty Constables and Tithingmen, or other Officers, to return Annual Militia Lists.

To the Chief Constable of the Division *of* Langbaurgh West *in the said County.*

North Riding of the **County of** York

BY VIRTUE of the several Acts of Parliament relating to the *Militia, WE, His Majesty's Lieutenancy* of the North Riding of the County of York assembled at a General Meeting held in and for the said Riding on Tuesday the 4th Day of January instant, do hereby direct you to issue out Orders, under your Hand, (according to the Form herewith sent), to the Constables, Tithingmen, or other Officers, of every Parish, Tithing, and Place, within your Jurisdiction, requiring them to return, upon Oath, to the Deputy Lieutenants and Justices of the Peace, at their first Meeting, to be held for the Subdivision of Langbaurgh West – at The National School Room in Stokesley in the said Riding, on Saturday – the 5th Day of February next, by eleven o'Clock in the fore noon of the same Day, fair and true Lists in Writing of ALL the Men usually, and at the Time when they shall make such Lists, dwelling within such Parishes, Tithings, or Places, between the Ages of Eighteen and Forty-five Years; And also requiring them to affix a true Copy of such Lists on the Door of the Church or Chapel belonging to their Parish, Tithing, or Place; and if any Place shall have no Church or Chapel belonging thereto, then on the Door of the Church or Chapel of some Parish or Place thereto adjoining, on some *Sunday* Morning before they shall make return of the same, which Sunday shall be *three days* at the least before such Meeting. And further requiring them to affix Notice in Writing, on such their respective Lists, of the Day and Hour, and Place of the said Subdivision Meeting for hearing Appeals under the said Acts. Herein fail not at your Peril. Given under our Hands and Seals the said fourth Day of January 1831.

[illegible signatures]

Geo. Strickland

Note.—The Chief Constable is required to insert the Day and Place of Meeting, and date and sign the Order, and im[mediately] give or leave the same with every Petty Constable or Tithingman, or other Officer, within his Hundred or Liberty, [wh]o will make out their Lists according to the Form herewith sent, after they shall have received the Notices from the Occupiers.

By the Act 42 G. III. c. 90, s. 28, 31, and 32. *If any Constable, Tithingman, or other Officer, shall refuse or neglect to appear, or return Lists, or shall be guilty of Fraud, Partiality, or gross Neglect, he shall be committed to Gaol, or be fined from 2l. to 20l.—Penalty of 50l. endeavouring to prevail on Constables, &c. to make a false Return—and of 10l. on Occupiers, for refusing to tell their Names, or the Names of Lodgers; and of 5l. for neglecting to deliver a correct and complete List.*

By the Act 46 G. III. c. 91, s. 4. The Ministers and Churchwardens, and Overseers of the Poor, and other Parochial Officers, in every Parish or Place, are required to aid and assist the Constables, Tithingmen, or other Officers, in making out the Lists, if necessary.

The Constable, Tithingman, or other Officer of any Parish, Tithing or Place, before they deliver the Notices Schedule A. must sign the same.

N. B. The Forms, Schedule A and B, can be procured from the Clerk of the Sub-division Meetings.

2—121. 1500. Printed by G. Duckworth, 76, Fleet Street, for His Majesty's Stationery Office. June, 1828.

An order dated 4th January 1831 to local officers of the North Riding, Yorkshire, to return to the authorities annual militia lists.

Captain Whinyates Royl Horse Artill

By His Majefty's Command.

Adjutant General's Office,
1st December, 1796.

RULES AND REGULATIONS

FOR THE

SWORD EXERCISE

OF THE

CAVALRY.

LONDON:
PRINTED for the WAR-OFFICE;

AND SOLD BY

T. EGERTON, MILITARY LIBRARY, WHITEHALL.

Title page of *Rules and Regulations for the Sword Exercise*, which bears the signature of Captain (Edward Charles) Whinyates, Royal Horse Artillery, who distinguished himself during the Napoleonic Wars. He later rose to the rank of Major General.

Pursuant to an Order from the Right Honorable Horatio Lord Viscount Nelson K.B. Duke of Bronte in Sicily, Vice Admiral of the Blue and Commander in Chief &c. &c. &c. — We whose Names are Hereunto Subscribed have been onboard His Majesty's Ship Edgar, and have there taken a strict and careful Survey on the Ordnance Stores remaining onboard and which we find to be as follows.

Remains June 1st 1801 —

Item		Size	Quantity
Iron Ordnance		32 Pound.	28
		18	28
		9	16
Ship Carriages		32 Pound.	28
		18	28
		9	16
Axeltrees, F. H.		32 Pound.	[illegible]
		18	[illegible]
		9	3.1
Trucks, Pairs, F. H.		32 Pound.	3.1
		18	4.1
		9	6.5
Beds and Coins, B. C.		32 Pound.	32.32
		18	32.32
		9	18.18
Coining Coins, N. P.			64
Shot	Round	32 Pound.	1520
		24	114
		18	1489
		9	685
	Double Headed	32 Pound.	84
		18	81
		9	47
	Grape	32 Pound	86
		18	162
		9	60
	Tin Case	18 Pound.	151
		9	112
Boxes for ditto Grenand Carcasses			85
Paper Cartridges		32 Pound.	1536
		18	1721
		9	850
Wadhooks		32 Pound.	1
		24	3
Aprons of Lead			20
Breechings		6½ Inch	12
		5½	14
			24
Tackles complete		2 Inch	21
		1½	—
Tacklefalls		Inch	
Carronade Locks			14
Boxes for Flannel Cartridges			3
Corn'd Powder		Pounds	27180
Copper hoops, on whole Barrels			1632
Tanned Hides			11
Ball Cartridges		Musket	7000
		Pistol	2000
Flannel Cartridges		18 Pound.	67
		9	101
Covers of Cured paper		18 Pound.	89
		9	107
Patent Staves		32 Pound.	2
		18	2
		9	1
Punches for Vents		18 Pound.	21
		9	13
Hammers for ditto			1

All which Stores we find as before specified, and do declare that we have taken this Survey with such Care and Equity, that we are ready (if required) to make Oath to the impartiality of our proceedings.

Given under our hands onboard His Majesty's Ship Edgar in Kioge Bay 17th June 1801 —

James Smith Defence
W. Gifford. Polyphemus
Henry [illegible] Agamemnon
} Gunners.

Details from a Naval Inventory of gunnery stores held on board H.M.S. *Edgar*, a 74 gun ship, on 1st June 1801. All stores are itemised, and the accuracy is vouched for by three gunners.

The Right Honourable Hugh Elliot GOVERNOR, and COMMANDER in CHIEF of the FORT and GARRISON of FORT ST. GEORGE, and TOWN of MADRASPATNAM, and of all the Forces which are, or shall be employed for the Service of the UNITED COMPANY of MERCHANTS of ENGLAND, Trading to the EAST INDIES, in the said FORT, GARRISON, and TOWN, and PRESIDENT of the Council of FORT ST. GEORGE, and the rest of the Council thereof.

TO *C. F. Tolfrey*

GREETING.

WE Reposing especial trust and Confidence in your Loyalty, Courage, and good conduct, do by these Presents, constitute and appoint you to be [illegible] *in the said United Company's Service, and do give and grant you full power, and authority to take your rank as* [illegible] *in the said Service from the* [illegible] *One Thousand Eight Hundred and* [illegible]

You are therefore carefully and diligently to discharge the duty thereof, by Exercising and well disciplining the inferior Officers, and Soldiers, who may, from time to time, be put under your Command, and we do hereby Command them to obey you as [illegible] *and you are to observe and follow all such orders and directions from time to time, as you shall receive from us, the Governor in Council, for the time being, or any other your Superior Officer, according to the rules and discipline of War. In pursuance of the trust hereby reposed in you; failing therein, these Presents to be annulled and void. Given under our hands, and the seal of the said Company in Fort St. George, this* [illegible] *Day of* [illegible] *in the* [illegible] *Year of the Reign of our Sovereign Lord GEORGE the Third, King by the GRACE of GOD, of the United Kingdoms of Great Britain and Ireland, and in the Year of Our Lord, One Thousand Eight Hundred and* [illegible]

ENTERED in the SECRETARY'S OFFICE, By order of the President in Council.

G. Strachey
Chief Sec[illegible]

[illegible]

Commission dated 27th June 1811 granting a Captain-Lieutenancy to Charles Frederick Tolfrey in the 24th Regiment of Native Infantry. While the Sovereign granted commissions to members of the Regular forces, until 1857 commissions in the Indian Army were granted by the Directors of the East India Company.

►

An open letter by Lt.-Col. Francis of Madras Cavalry pressing for the appointment of a Colonel for each regiment of cavalry. Again, prior to 1857, it was the Directors of the East India Company who were responsible for the greater part of the armed forces in India.

TO

THE HONOURABLE THE COURT OF DIRECTORS,

&c. &c. &c.

Honourable Sirs :

Private letters from Madras mention that it was the intention of Colonel Floyer to submit to your Honourable Court how disadvantageous the existing Regulations are to your Cavalry; but, as accounts have been received of that officer's death, and being apprehensive that severe indisposition may have prevented the fulfilment of his intention, I feel confident your Honourable Court will pardon the liberty I have taken in addressing you on a subject which I beg leave to consider of the utmost importance to that branch of your army in which I have served twenty-nine years.

Since the Regulations of 1796, your Honourable Court deemed it proper to issue orders, in consequence, I believe, of the subject having been brought to your notice by individuals, directing that a colonel should be placed at the head of *each battalion* of Artillery. The corps of Engineers also have had a colonel granted them since that period; and when once the minds of your Honourable Court are entirely given to the present situation of your Cavalry, I humbly conceive you will see the expediency of giving a colonel to each regiment, to keep up the respectability of that useful and distinguished branch of your service, and to afford it an equal chance with other corps for that promotion which constitutes the life of every army in the world.

Your Honourable Court may fairly boast of having in your service a body of Cavalry as highly disciplined and efficient as can be produced by any other country in the world, and yet every other country considers a colonel to each regiment indispensably necessary.

Several of your cornets are of thirteen years standing, and I believe the case of Cornet Logan was not long ago submitted to your Honourable Court praying for the brevet rank of *lieutenant*, that he might avail himself of the brevet rank of captain when he had served fifteen years, as there was little hope of his promotion to the rank of lieutenant by regimental rise within that period.

Some of your captains of Cavalry are twenty-five years in the service, and, in general, your officers cannot expect to attain that rank in their regiment in less than twenty-two years. Major Foulis is thirty-two years in your service, without any near prospect of a lieutenant-coloneley, and I believe Lieut.-Colonel Cosby, was thirty-five years in the service before his promotion to that rank.

I presume that your Honourable Court will concur in the opinion that a Cavalry corps should not be left in the charge of junior officers, but be commanded by their respective lieutenant-colonels, who, from their long experience and knowledge of the service, have the requisite influence over all ranks. But if these officers are worn out before they attain the rank of lieutenant-colonel, they become scarcely able to conduct a regiment through its common drills and field-days, much less to lead it into action with that ardour and energetic dash so requisite on such an occasion. An old officer thus worn out is frequently so much respected by his brother-officers that they shrink from the heartrending task of applying for his being invalided, valuing him, like the old hound in the fable, for past, faithful, and meritorious services; the corps, however, notwithstanding this proper feeling, suffers in its efficiency.

When the present augmented strength of a regiment of Cavalry is taken into comparative consideration with a battalion of Artillery and the corps of Engineers, I trust it will be the opinion of your Honourable Court that in point of numbers and responsibility a regiment of Cavalry is equally well entitled to a colonel. The consideration of the subject also with respect to the amount of off-reckonings contributed to the Off-reckoning Fund will, I presume, appear to your Honourable Court to bear investigation.

The strength of a regiment is as follows, *viz.*

Lieutenant-Colonel - -	1	
Majors - - - - - - -	2	
Captains - - - - - -	3	
Lieutenants - - - - -	8	
Cornets - - - - - -	6	
Subadars - - - - - -	8	
Jemadars - - - - - -	8	
Havildars - - - - -	35	
Naigues - - - - -	33	
Trumpeters - - - - -	8	
Farriers - - - - -	8	
Native Doctors - - -	2	
Troopers - - - - -	640	
		762
Horses - - - - - -	742	
Grasscutters - - - - -	742	
Horsekeepers - - - -	262	
		1,746
Total - -		2,508

N.B. Not including the Horses, Horsekeepers, and Grasscutters of European officers.

If the six-pounder Gallopers and detachment of Horse Artillery still form a component part of a corps of Cavalry, it will add to the already enumerated numbers and responsibility.

The recent increase of two regiments of Infantry at Madras has given the Colonelcy and Offreckonings to all officers above Colonel Marriott, who came into the service as a Cadet eight or ten years after the following officers, *viz.* Major-general Ramley, Major-general Doveton, Colonel Nuthall.

The same observation may be made relative to the Senior Lieutenant-Colonels of Cavalry on the Bengal Establishment; and I feel confident that, on a comparative consideration of rank, your Honourable Court will be of opinion that promotion ought to be extended to the Bengal and Madras Cavalry, thus rendering the present proposal in that point of view unobjectionable. I can safely assure your Honourable Court, that I have only expressed the sentiments of every Cavalry officer in your service, and that your granting them a Colonel to each regiment will be considered by them as removing the present existing distinction, as marking the estimation in which you hold their services, and showing your determination not to withhold any means in your power to forward fair promotion, to which every officer naturally looks up to support him in the active and zealous performance of his duty in a climate bearing so very hard on his constitution.

Trusting your Honourable Court will impute this address entirely to an anxious desire that you may be induced to turn your attention to the subject, I have the honour to be,

Honourable Sirs,

Your very obedient and very faithful humble servant,

December 1, 1819.

FRANCIS A. DANIELL.

Lieutenant-Colonel Madras Cavalry.

PASS. V. Scots R. Greys. B TROOP.

Preston Barracks Brighton 26th Dec 1874

The bearer hereof No. 1424 Pte R Parker

has leave from 6 a.m. the Inst till 10 p.m.
the 28th Instant for London

Approved C F Watkins [illegible] Commanding Troop.

Granted [illegible] Colonel Commanding Officer
Scots Greys

Leave pass for a member of B Troop Scots Greys granted December 1874.

A Captain's commission issued by George III on 18th June 1801.

George R

George the Third by the Grace of God of the United Kingdom of Great Britain and Ireland King Defender of the Faith &c. To Our Trusty and Welbeloved Thomas Harrison Gent Greeting: We do by these Presents Constitute and Appoint you to be Lieutenant to that Troop whereof [illegible] Esq.'s Captain in the Royal Westminster Volunteer Cavalry commanded by Our Trusty & Welbeloved Major John Elliott but not to take rank in Our Army except during the time of the said Corps being called out into actual Service. You are therefore carefully and diligently to Discharge the Duty of Lieutenant by Exercising and Well disciplining both the inferior Officers and Soldiers of that Troop and We do hereby Command them to Obey you as their Lieutenant and You are to observe and follow such Orders and Directions from Time to Time as you shall receive from your Major Commandant or any other your Superior Officer according to the Rules and Discipline of War in pursuance of the Trust hereby reposed in You. Given at Our Court at Saint James's the eighteenth day of June 1801 in the Forty first Year of Our Reign.

By His Majesty's Command

Pelham

[illegible] with the [illegible] at War [illegible]

[illegible] Harrison Gent. Lieut. in the Royal [illegible]inster Volunteer Cavalry

Ent with the Compy General of Musters
Wm Woodman

Victoria R.I.

Victoria by the Grace of God of the United Kingdom of Great Britain and Ireland Queen, Defender of the Faith, Empress of India and Sovereign of the Most Honorable Order of the Bath,

To Our trusty and well-beloved Warren Frederick Caborne, Esquire, Commander in Our Royal Naval Reserve,

Greeting. Whereas We have thought fit to nominate and appoint you to be an Additional Member of the Civil division of the Third Class, or Companions of Our said Most Honorable Order of the Bath, We do by these Presents grant unto you the dignity of a Companion of Our said Order, and hereby authorise you to have

Grant of the dignity of a Companion (Civil division) of the Order of the Bath to Warren Frederick Caborne, Esquire.

hold and enjoy the said dignity and rank
as an Additional Member of the Civil
division of the Third Class or Companions
of Our aforesaid Order, together with all
and singular the privileges thereunto
belonging or appertaining.

Given at Our Court at Windsor
under Our Sign Manual and the Seal
of Our said Order, this twenty second day
of June 1897 in the Sixty first
year of Our Reign.

By the Sovereign's Command.

Albert Edward P.
G. M.

Order making Warren Frederick Caborne a member of the Order of the Bath on 22nd June 1897. At the top can be seen the signature of Queen Victoria. Although the Order was a very ancient one, it did in fact lapse for a while and was not revived until 1725.

PLAYER'S CIGARETTES.
INFANTRY PRIVATE'S SHAKO, 1800-1806.
HIGHLAND REGIMENTS; Full dress bonnet, 1790.
43RD REGIMENT OF FOOT; Private's forage cap, 1852.
73RD REGIMENT OF FOOT; Officer's full dress shako, 1840-48.
6TH DRAGOON GUARDS; Officer's full dress helmet, 1842-47.
15TH THE KING'S HUSSARS; Officer's full dress busby, 1861.
4TH REGIMENT OF FOOT; Officer's forage cap, 1852-1861.
ISSUED BY JOHN PLAYER & SONS
BRANCH OF THE IMPERIAL TOBACCO Cº (OF GREAT BRITAIN & IRELAND), LTD.

Military Headdresses – a set of cigarette cards first issued by John Player & Sons, March 1931. The set of 50 covers the whole period from the 17th to the 19th century. Each carries on the back a generally accurate account of the headdress.

History of British Uniforms – a series of 50 cigarette cards issued by Carreras in 1937, again neatly done in good clear colours. Minor details are not always absolutely accurate, but on the whole the standards are high.

HISTORY OF ARMY UNIFORMS
A SERIES OF 50
No. 44
The 65th Foot (1746). The York and Lancaster Regiment.
In Army Lists for 1746-8 Sir William Shirley's American Provincials are noted as the 65th Foot, and picture shows a foot soldier of this corps, which was finally disbanded as the 50th Foot. Historically, the title belongs to the 1st York and Lancaster Regiment which was raised in 1756, and became a distinct corps with the title of 65th Foot in 1758. County titles were conferred in 1782, the regiment was then styled the 2nd Yorkshire, North Riding Regiment. The badge of the Royal Ti er was conferred on the 65th for distinguished conduct in India.
CARRERAS LTD
(ESTD 1788)
WORKS, LONDON, ENGLAND
HISTORY OF ARMY UNIFORMS
A SERIES OF 50
No. 45
The 69th (South Lincolnshire)

THE PRINCE OF WALES'S VOLUNTEERS
(Sth. Lancashire Regt.) 40th & 82nd Foot.

66 B.D.V. CIGARETTES

THE NORTHAMPTONSHIRE REGIMENT
48th & 58th Foot.

65 B.D.V. CIGARETTES

ORDER OF ST. MICHAËL AND
ST. GEORGE (GRAND CROSS)
(GREAT BRITAIN)

Series 10—No. 12 B.D.V. CIGARETTES.

ARMY SERVICE CORPS.

84 B.D.V. CIGARETTES

Example of Victorian scrapbook material. The figure of the cavalryman with silver helmet and blue uniform is slightly embossed and in bright colours. These scrapbooks were very popular, and a number of military subjects were included in the selection.

◀

Cards from a series of "Crests and Badges of the British Army", issued by B.D.V. Cigarettes between 1910 and 1925; the complete set numbering 108. The pictures are not embroidered or woven, but instead are printed on the silk.

Silver whistle and chain from officer's cross belt. They bear the assay marks and date letter for Birmingham 1903. The lion's head was secured to the belt, and farther along was the receptacle shown in the centre of the picture; into this fitted the whistle; held by a springclip.

Example of a Staffordshire china commemoration of a war hero – in this case Lord Kitchener, First World War. It stands some 5 in. high.

Brass matchbox holders, or perhaps Bible clips, made from sheet brass and the centre dome from First World War German belt buckles. They are typical examples of "trench art", that is, commemorative material made from various war items.

Souvenir badge of the 32nd Encampment of 1898 of the Grand Army of the Republic, specially made for this reunion of American Civil War Veterans. It was manufactured by G. M. Robbins of Attleboro, Massachusetts.

A miniature tableau set in a normal brandy glass representing Christmas in the Peninsula 1812. Manufactured by S. J. Whitehead.

V

MILITARY MINIATURES

Collectors of military miniatures – toy soldiers to the less sophisticated – are of three main groups. There are those whose interest lies acquiring early examples of the craft; those interested in modern, detailed miniatures and those whose concern is with war games or *kriegspiel*.

Each group can trace their common interest as far back as Ancient Egypt, although it is unlikely that any collector will ever be lucky enough to acquire any examples dating back farther than the 18th century. With very early examples it is impossible to be precise as to their purpose, and they may well have been either funerary objects or toys. In the case of the Egyptian examples, fashioned in wood, they were almost certainly funerary, but with those of Greek and Roman times it seems far more likely that their prime object was that of play. Some of the surviving examples are of gladiators, others are of horsemen and legionaries.

By the early Middle Ages puppet-like models of knights are illustrated in some manuscripts, and by the 16th century these had developed into superb models of mounted knights. In imitation of the true tilt these models were pushed or pulled along in an attempt to dislodge the opponent's figure by striking it with a lance fixed beneath the model's arm. The horses were usually hollow, but the detachable knight was cast in solid bronze. Several examples of these charming figures have survived and may be seen in various museums.

Model soldiers have been produced in a great variety of materials, cardboard, pottery, wood, lead and precious metals are all mentioned in contemporary accounts. Louis XIII of France is known to have possessed a small army of miniatures, including machines of war, and this collection was increased by a gift from his mother of 300 silver soldiers. Louis XIV also had a similar miniature army, and later he ordered a very fine set for his son. Many of these figures are described as being fitted with automatic mechanisms which enabled them to perform certain military manoeuvres. Perhaps the cost of these fine figures had some bearing on his later order, for these were of cardboard.

In the 17th century Germany was already acquiring a tradition as being the source of good-quality model soldiers, and by the 18th century the craft had become a flourishing industry. Most of the early examples are crude, lacking any fine detail and were of the type known as "flats" since they are only two-dimensional, with little attempt to suggest solidity of contour. Moulds were fashioned from two pieces of slate on to which were engraved the shapes. The two blocks were then clamped together and an alloy of tin and lead poured in, the excess metal draining off through escape channels.

It was in Nuremburg, already famous for its armour, that the industry began to flourish under the guidance of members of the famous Hilpert family. Soon they were producing a wide range of figures, mostly about two inches high. Early examples were made from almost pure tin, but although tin's softness allowed good-quality engraving to be carried out, the figures were very soft, and consequently very liable to damage. Hardness was increased by adding more lead to the tin, and the resulting figures were much firmer. Many other makers, stimulated by Hilpert's success, also began producing these flats, and a number of their 19th-century examples have survived.

Almost the entire German production was of flats, and it appears to have been in France that the first large-scale manufacture of solids was undertaken. It also seems most likely that the military glory of Napoleon was responsible for this development. The figures were three-dimensional, solid and cast with a mixture of lead and antimony; they stood approximately two inches high. Some were cast in one piece, while others, the better-quality ones, were assembled from separate pieces – arms, head, torso, legs and accessories. In Germany the large-scale production of solids did not develop until a much later date, although some had been manufactured there for many years. One firm, Georg Heyde, of Dresden, manufactured an enormous range of figures, extending from the Ancient World to early 20th-century India.

For many years the great majority of toy soldiers used in England were imported from Germany and France, and there is little evidence to suggest any home production in quantity. There were presumably some resources available early in the 19th century, for a Captain William Siborne was commissioned to construct an accurate model of the Battle of Waterloo, and his finished effort, completed after some twenty years, incorporated about 190,000 half-inch figures. He then set to work to produce a second and better model, and in this one his figures had movable arms and innumerable detailed pieces of equipment. Even this less massive model used some 3,000 figures, and although it is not certain, it does seem likely that Siborne made most of them himself.

One serious, limiting factor in the marketing of solid models was the cost, for the amount of material used was considerable. It was a maker of mechanical toys who made the breakthrough. William Britain used an alloy in place of the more expensive lead, and in 1893 was able to hollow-cast the figures so that they were now a mere shell of metal. Weight and material were reduced, resulting in a cheaper product which still possessed the virtues of the German solid cast models. Unfortunately for Britain there was little demand until a well-known London department store mounted an impressive display of his wares.

Interest and demand were stimulated, and soon William Britain was producing a wide range of figures featuring many races and armies. So successful did his figures become that soon they were being exported to the home of the solid, Germany.

Naturally Britain's first lines featured well-known British regiments, and the Life Guard and The Black Watch were among these. As demand increased, his range was expanded and came to include Turks, Italians, Egyptians and Austrians – well over one hundred different lines in all. Soon he was making artillery, tanks and similar war equipment. Noting his success, some competitors sought to pirate his designs, and from 1900 all his figures bore, somewhere on their stands or anatomy, the words claiming his copyright. Early examples of Britain's products, especially groups such as the Royal Horse Artillery battery, are now much sought-after by collectors. Most of Britain's soldiers were sold boxed in sets, and the familiar shiny red cardboard boxes may still be found in odd places.

Hollow-cast figures suffered from one serious handicap: they were rather fragile, and many heads and limbs were restored with the aid of a broken matchstick. However, the advent of plastics solved this problem, and after the Second World War many firms throughout Europe turned to this new material, which gradually ousted the more traditional lead alloys. The majority of plastic figures are light, far less fragile and quite well detailed. A recent development in the field of plastic miniatures has been the production of very good-quality models of armours in the collections of H.M. Tower of London and the Royal Armoury, Madrid. These are far more expensive than the usual commercial product.

One interesting development in the field of plastic models has been the production of interchangeable parts which enable the basic figure to be varied by changing the head, torso and accessories.

For the purist, seeking far greater accuracy and detail, there are the specialist's models, and these are normally solid-cast in various alloys. In general, they are of the 54-mm size, although there are variants. Quality is extremely high, and the detail is usually superb, although obviously there is some slight variation according to price range – those in the cheaper bracket being just that little bit less finely modelled. Most of these specialist models can be purchased either blank or painted, and naturally the price reflects the time and trouble required, for a painted figure may well cost five or six times that of a blank.

For most collectors it is the blank that will attract, for this affords the very considerable satisfaction and pleasure of painting it and bringing it to life. There is no point in minimising the difficulty involved in the painting, for it is a task that requires concentration, skill, patience and research – above all, it is a job that takes up a great deal of time. Some of these quality blanks are sold complete with full painting instructions for the uniform, equipment and accessories, but if this is not the case, then research must be undertaken. Such knowledge is essential before undertaking any painting or preparation, but a growing demand has stimulated an increasing flow of reliable textbooks which readily supply most details.

First step must be the cleaning up of the blank, for in all but the very best there are almost certain to be thin ridges left by the joints of the mould, and these must be carefully trimmed and smoothed away, as must be any air bubbles or gaps. Now a base is required, and the composition will vary according to the alloy used and paints recommended. Usually a flat coat of thinned white paint is sufficient, but many makers recommend special preparations.

Final painting is a delicate task requiring concentration, pertinacity and strict attention to detail. Two alternatives are open to the collector, for either oil or water colours can be used, and choice is largely a matter of personal preference. Before painting, most collectors prefer to mount the figure on some form of handling block, for all colours will rub if handled too frequently.

Sequence of painting will frequently be determined by the pose of the figure and convenience of access. Faces are extremely important, and several fine brushes trimmed almost to nothing will prove most useful in coping with eyes and facial features. Eye strain and tedium can both be alleviated by the use of a lens mounted on an adjustable stand, and a watchmaker's glass is useful. The real skill lies in the shading and creation of facial modellings, for it is such detail that will bring the model to life. Uniform details are of supreme importance, and obviously this information must be ready to hand for reference.

Although the majority of these figures are cast in alloy, there is at least one exception, for the French series, Historex, are finely cast in plastic. This form of production offers certain advantages, for it means that many of the minute detailed pieces, such as lanyards and belts, are cast separately and can therefore be painted easily, since there is no fear of smudging adjacent parts of the model. Once the piece is dry, it can be stuck in position, although this is a task requiring care and application if the plastic surface is not to be marked. Again, this casting of separate pieces enables variations on a basic figure, and many producers of metal castings offer the same facility, having a basic torso to which may be added a variety of heads and limbs.

Many collectors specialise in their own conversion of models, and modify both alloy and plastic figures by the judicious addition of pieces and alteration of limbs. This offers fascinating possibilities, and modern adhesives have made it easy to change a figure almost beyond recognition. Foil from toothpaste tubes makes extremely effective material which can be used to alter coats and hats, while judicious bending and soldering can alter the position of arms and legs. The possibilities are enormous, and the interested reader is recommended to study the relevant chapters of the books listed in the bibliography.

Attractive display is essential if these figures are to be seen to full advantage, and there will be few that deny they should be under glass. Dust, fingering and chance knocks are all minimised, and if some form of lighting is fitted inside the glass case the effect is extremely attractive. Background is a matter of personal taste – some collectors seek to form associated groups of figures and build up small dioramas complete with scenic effects. Others simply have the figures free-standing, either on the metal plate or fitted to a block of wood which forms a plinth.

War Games are a specialised feature of model soldiers that has grown tremendously in popularity over the past few years. Essentially it is a dice game, but one depends not primarily on chance but on skill and planning as well. Although models have been used for centuries when planning battles, campaigns or simply practising, the credit for war games as played today can probably be allocated to H. G. Wells. In 1911 he published a small book entitled *Floor Games*, followed in 1913 by *Little Wars*, and in both books he developed the idea of creating "Toy lands" in which to have adventures and act out stories. He advocated, with enthusiasm and humour, the firing of toy cannon against armies of toy soldiers, and explained how battles developed, were waged and lost or won. His infectious enthusiasm spread slowly, but never died, and after the Second World War burst out once again into full glory.

Few enthusiasts would now tolerate the physical firing of toy cannon against their armies, but the effect of cannon fire is simulated. There is no one set of rules for war games, but rather a series of rules for every type of combat. Essentially the troops are set out on a terrain made realistic by models and paintings, and the object of the campaign is agreed on in advance – the taking of a village, the turning of a defence line or just a set battle. Movement of the troops is strictly governed by dice and tape measure, as are casualities and objectives. There is little point in attempting to outline rules here, for the ramifications are enormous, and the reader is referred to the bibliography for suitable sources.

Figures used in war games are generally small, for obviously space is an important consideration, and an army of fifty 54-mm figures requires a fair amount of space as well as a considerable amount of money. Most popular for this pastime are those measuring 20 mm, which is the usual 00 gauge railway size, or there are slightly larger 30-mm figures, as well as some of 40-mm size – although here, size and cost again begin to limit the size of armies.

Expense is obviously a prime factor, and there is little doubt that plastic figures offer one of the cheapest ways of building up an army at low cost. Airfix 00 scale figures are ideal and offer a variety of figures from Romans to Second World War Russians. A box contains between twenty and forty pieces, and an army of a hundred or more can be built up for a very small amount. Another advantage is the availability of a number of buildings, vehicles and other accessories, all to the same scale. Each figure is accurately and well cast, and there are a variety of poses in every group.

More costly and frequently slightly larger are the metal flats, and these are produced in many countries – again in a tremendous range, including Ancient Egyptians, Saxons, Vikings, Normans and Burgundians.

Painting the large number of figures needed for war games is obviously a very different proposition from that involved in large-scale figures, and most enthusiasts seek to give a general effect rather than detail. Face and hands are frequently left as flat washes of paint, as are jackets and trousers. Since the figures need to be handled, it is as well to apply a coat of clear varnish to the completed figure.

For the collector of "antique" model soldiers the sources are likely to be general antique shops, for there is a growing awareness of their value. It is always worth looking in at jumble sales and similar functions, for many boyhood treasures are disposed of there, and many a cherished box of Britain's soldiers has been found at a church sale. Specialist figures are increasingly available from a number of sources, and most manufacturers issue some form of catalogue. Most of the specialist magazines and societies carry details of these suppliers, as do some of the relevant books.

For the collector of militaria with limited finance and space the model soldier offers a very satisfying scope and one which is far more active than any other form of collecting.

Superbly painted figure of an officer of the 1st Polish Lancers of Napoleon's Imperial Guard. French model, in plastic, in which each item is cast separately.

Left: A flat of the Prussian Hussar General Von Werner, by Ochel. It hardly appears to be flat, it is so well painted.
Right: A very nicely done model, 20 mm, of a Saxon Uhlan about 1900. The man may be dismounted and will stand on his own.

Left: A Landsknecht figure by Imrie-Risley, a U.S. model maker. *Centre:* Spanish figure of an officer of the Palace Guard about 1715, by José Almiral. *Right:* Another high-quality figure by Stadden showing an officer of the Second Dutch Lancers of Napoleon's Imperial Guard.

A small Stadden figure of a Grenadier of the Imperial Guard.

Left: A German Panzer officer, made in the U.S.A. *Right:* A Grenadier of the Second Empire, another figure cast in plastic.

Below, left: Officer of the 6th Inniskilling Dragoons of 1797. *Centre:* Mass-produced plastic figure of a German soldier of the Secon World War. *Right:* S.S. Division Leidstandarte officer, U.S.A. made.

A British Lancer Trooper *c.* 1910. This figure has been converted from a Britain mounted figure.

An Adjutant in the Prussian Life Guard Hussars, *c.* 1912; a Stadden figure.

Semi-flats of the 1920s. The metal is of poor quality with a high content of lead. The detail, casting and painting are crude.

Left: A French Knight De Cordeboeuf done by the famous modeller Courtney. *Centre:* A mounted Norman knight, one of the "Willie" figures made by Captain Suren. *Right:* Richard I by Ping.

A group of crude solid figures, again of the 1920s and probably of British origin. Detail is either sketchy or non-existent.

Left: A well-modelled figure of a Gaulish trumpeter by Hinton Hunt.
Centre: a Gaulish chieftain, this time by Russell Gammage.
Right: Edward III – a finely detailed figure by Stadden. Although all models of this calibre are well made and accurate in details, the finishing touch lies, of course, in careful painting.

Selection of scale plastic figures by Airfix Ltd., produced for model railways and used also by War Gamers. *Left to Right:* German Infantry, French Bugler, British Infantry; all First World War.

German *Zinnfiguren,* Tin Soldiers, flats of warriors of the Dark Ages.

Additional figures by Airfix Ltd. *Left to right:* American Infantry (2), First World War, Roman Standard Bearer and Roman Centurion.

German armoured car Sd Kfz 234, another product of Airfix Ltd. The model has rotating wheels; the gun and figures are movable.

The coin on the left, diameter $1\frac{1}{8}$ in., indicates the size of these 20- and 30-mm figures. *Left:* German Infantry, First World War, by Greenwood and Ball. *Centre:* Prussian Infantry, *c.* 1900, by J. Scruby U.S.A. *Right:* Bavarian Infantry, *c.* 1900, by Greenwood and Ball.

"Peeler" or early London Policeman, *c.* 1830.

Left: Cromwellian officer with foaming tankard by Les Higgins. *Centre:* Caucasian rifleman, *c.* 1855 by Neville Dickinson. *Right:* Early 30-mm Stadden figure of a German First World War Infantryman.

Bow Street Runner, a civilian forerunner of the 18th-century policeman.

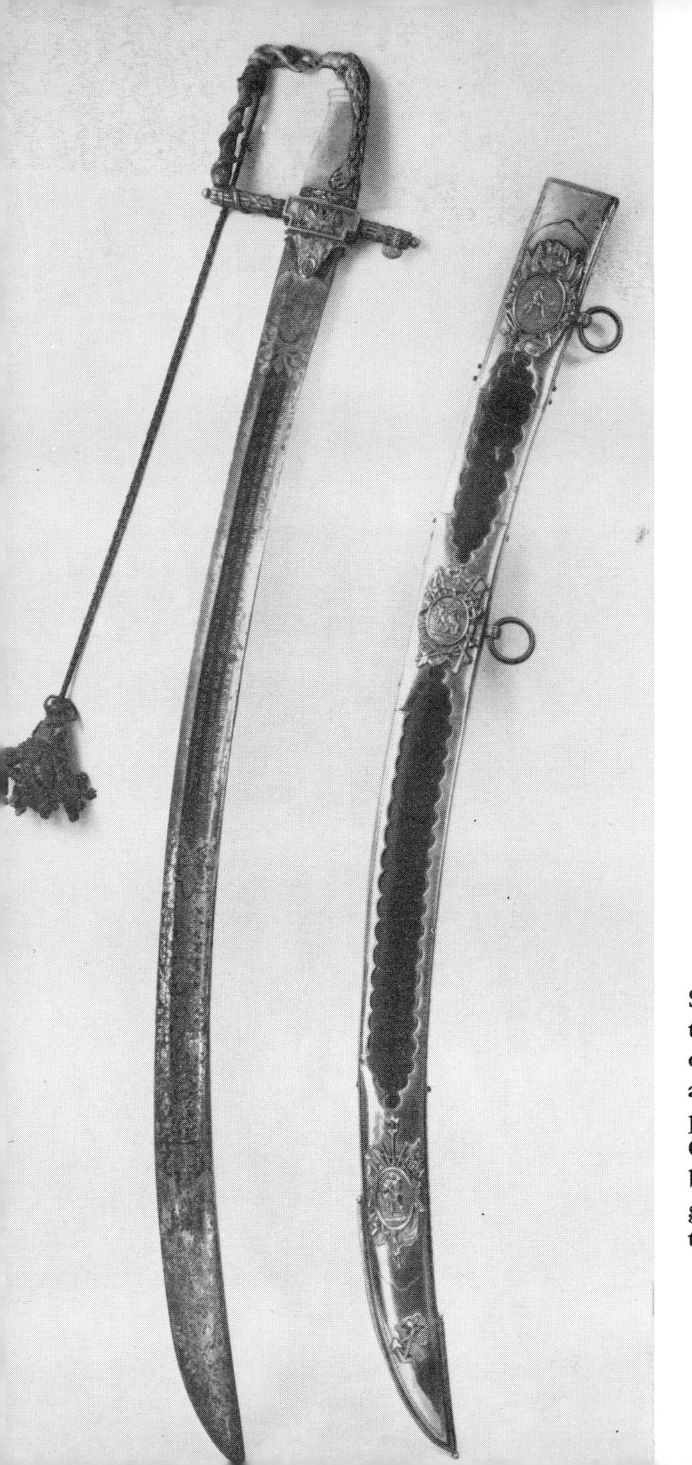

Sword presented by Lloy to the Hon. G. A. Crofto of H.M.S. *Canbrian* for h action against a Spanish privateer; September 180 Ornately decorated, the blade bears an inscriptio giving various details of this particular action.

VI

MILITARY WEAPONS

Obviously in times of need any weapon becomes "military", but in collecting circles the term is understood to apply to weapons used by the regular armed forces of any country. In fact, the line of distinction between these and non-military is sometimes rather vague, but in the great majority of cases the weapon bears some marks indicating that it was supplied by the authorities or else is in conformity to a pattern laid down by such authorities. Frequently such marks take the form of a royal cypher or perhaps the mark of a national arsenal. The majority of earlier British military firearms carry the royal cypher as well as the name of the manufacturer; others bear the royal cypher and the word "Tower". Many of the martial weapons of the United States bear the name of an arsenal such as Springfield or Harpers Ferry. With the later advent of mechanisation and mass production, simpler identification was achieved, in the case of British weapons by an arrow mark and the letters B.O. (Board of Ordnance) or W.D. (War Department). However, in the case of swords the position is less clear, for frequently they were purchased from the manufacturer either in bulk or individually by officers, and so frequently lack marks specifying a military use, although most conform to a pattern.

In practice, military weapons within the above definition are limited to the period from the 17th century onwards. Prior to this period equipment had been very largely a matter of personal choice and cost. A few of the finer weapons are marked with the arms of some ruler or other, but most cannot, with any certainty, be assigned to a particular unit.

From the 17th century onwards many countries began to adopt standard patterns of military weapons, and in England the Council of War in 1630 was specifying the overall and barrel length of military firearms. Pistols were to be 26 inches overall, with an 18-inch barrel, while muskets were to be 62 inches overall and have a 48-inch barrel. Pistols were fitted with that type of ignition mechanism known as the wheel-lock, while muskets utilised the simpler and cheaper matchlock or snaphaunce system of ignition.

During the reign of Queen Anne (1702–14) the British Army's last remaining matchlocks were replaced by the flintlock, a mechanism that was to give excellent service until around the middle of the 19th century. Military flintlock weapons were mostly assembled at the Tower of London from stocks of brass and iron fittings held there in store. Barrels and the locks were generally made in Birmingham, where they were examined by an Ordnance viewer, tested and if satisfactory sent on to the Tower. Some of the actual assembly was undertaken in the Tower itself, for there was a small factory within the walls, but in times of heavy demand the basic components were supplied to local gunmakers – and there were many living in the vicinity of the Tower – who then assembled the weapons.

During the 1720s there first appeared that most famous musket the "Brown Bess". This long arm with its pleasing lines was originally fitted with a 46-inch barrel and wooden ramrod, and was known officially as the Long Land model.

A very similar weapon fitted with a 42-inch barrel was known as the Short Land model. The locks of the Brown Bess originally carried the maker's name and date of manufacture engraved thereon, but in 1764 this practice was abandoned and the word "Tower" substituted.

In 1768 the old Long Land Model with its 46-inch barrel was superseded as the standard army weapon by the Short Land Model, although some of the 46-inch models were still being manufactured as late as 1790.

The outbreak of the French Revolutionary Wars found the Ordnance – the department of the Army responsible for the supply of equipment – very short of weapons, and to remedy the deficiency a large number of muskets were purchased from the East India Company. The India pattern was very similar in general design, but was fitted with a 39-inch barrel. In 1802 there went into production a 42-inch barrel weapon, known as the New Land Model, which differed in certain respects from its predecessor but was to continue in service largely unchanged until the percussion system was adopted in 1838. From 1838 onwards the changes in design and models of the British Army's muskets and rifles were considerable. Most models adopted were criticised on one count or another until the Short Magazine Lee Enfield was adopted, and this was to give steady service in both World Wars until this in turn was ousted by the present semi-automatic models.

Most British military firearms have stamped on them a crown and royal cypher which may range from C.R. (Charles II) to G.R. George VI), and in many cases the word "Tower". However, during the Napoleonic period vast numbers of volunteer groups of cavalry, infantry and artillery were formed. In some cases they were issued with weapons by the Government, but more frequently the volunteers supplied their own. Thus it is not uncommon to find a military type weapon bearing civilian names and lacking the royal cypher and Tower. Many of these volunteer weapons were marked with the County's initial and V for volunteer, frequently on the heel of the butt cap. A similar pattern occurs in the weapons of such countries as France and the United States. In France certain models normally designated by the year of their adoption were supplied to the French Army. In the United States after the American War of Independence two national arsenals were established at Harpers Ferry and Springfield, although the work of manufacturing the weapons was, in the early days, given to civilian contractors – Simeon North was probably the best known. There is a marked similarity in style between the early military firearms of France and the United States.

Most of the infantry muskets were intended to be used with bayonets, and in the 17th century these were of the plug type. They had a small tapering wooden grip which was pushed home into the muzzle of the empty musket to make what was in effect a short pike. Blades were normally short and fairly broad at the base. The obvious limitations of such a system stimulated designers to seek some means of fixing the bayonet while still allowing the musket to be fired. Eventually they produced the socket bayonet, which allowed the user to fit the bayonet to the outside of the muzzle, which meant that the soldier could still fire even with the bayonet in position. This method of attaching bayonets was commonly used until the mid-19th century, when a stud beneath the barrel and a slot and spring on the bayonet hilt became more common.

There was a general but gradual lengthening of the blade until the late 19th century, when the trend was reversed and the emphasis was on a much shorter blade. One reason for this changing emphasis was the diminishing use of cavalry, for the bayonet had been primarily an anti-cavalry weapon. Most modern bayonets have a fairly short blade. The available varieties of bayonets are enormous, and they offer great opportunities for the collector with limited resources. The growing interest over the last few years has resulted in a number of authoritative publications which will enable the collector to identify most specimens. The demand has forced up some prices, but there are still plenty available for very reasonable sums.

Besides the bayonet, most infantrymen of the 17th century also carried a sword or hanger. This was not discarded by the British until around the middle of the 18th century, and in many

European countries such as Russia the practice continued until very much later. Many of the early British military infantry swords were fitted with brass hilts, and the style is usually identified by a date. Unfortunately little is known for certain about the precise style, since the dress regulations of the time are either unknown or not very specific. The so-called 1751 pattern is, in fact, so dated from a series of pictures showing soldiers with this type of weapon, but may well have been in use earlier. After the latter part of the 18th century swords were, in general, used only by the cavalry and officers. Cavalry swords were of two main types: the cutting, slashing weapon with its blade curved to varying degrees, or the thrusting weapon with a long slim straight blade. Both types are used by cavalry all over the world, but many are identifiable by the markings which may be found on blade or hilt.

Pistols were essentially a cavalry weapon, although some infantry officers also carried them. They were usually carried in pairs in holsters at the horse's neck, a practice surviving until the introduction of revolvers rendered the duplication unnecessary. Most of those in use during the early part of the 17th century were fitted with the wheel-lock mechanism, but from the middle of the century onwards this was largely replaced by the flintlock. Most pistols had long barrels, normally 9–12 inches in length, and fired a ball around half an inch in diameter. Basic design of British pistols altered very little throughout the 18th century, although the introduction of the percussion system brought about a simpler shape lacking any trace of the slight swelling butt found on all previous models. Many of these pistols and the shorter musket or carbine used by the cavalry, as well as some muskets, were marked on the barrel with the initials of the particular regiment to which it was issued.

Officers frequently carried pistols, but these were generally civilian models which they purchased privately. Similar circumstances applied to the purchase of swords by officers, for many had weapons which were basically the official pattern but with variations in detail – perhaps the blade was engraved or decorated to personal specifications, or the chiselling on the hilt is of a better quality than most. The rules governing swords were set out in the Army Dress Regulations, which describe the uniform and weapons, but unfortunately the descriptions are not always as precise as one might like – thus those for 1796 state that the sword shall have a brass guard, pommel, shell gilt with gold and the grip or handle of silver twisted wire. The Dress Regulations of 1822 introduced for the first time for Infantry Officers of the British Army the so-called Gothic Hilt with its half basket fashioned from a series of bars with the royal cypher incorporated. This style, with modifications, remained in general use for infantry officers until 1892, when the guard of bars was replaced by a sheet-steel type still carried at present.

Armour is essentially military in character, for despite the ingenuity and skill of the armourer, it was never a garment for general wear, apart from war and parades. Although armour of one form or another has been in use since earliest times, for all practical purposes collectors cannot hope for anything much predating the late 16th century. By the beginning of the 17th century armour was being discarded by most units, and by the time of the English Civil War it was the exception rather than the rule for most troops. Pikemen, whose job was to guard the musketeers, wore a wide-brimmed helmet called a pot together with a back- and breast-plate fitted with a two-piece skirt. Cavalry wore a breast- and back-plate, a bridle gauntlet which extended to the elbow and on the head a burgonet. This type of helmet had a skull with a number of radial flutes, a flat peak pierced by an adjustable bar, the nasal and a flaring neckguard. Burgonets of this type were in common use in Northern Europe, and the English version had a pivoted peak fitted with a three-bar faceguard.

Infantry soon discarded their helmets, replacing them with wide-brimmed hats, while the cavalry soon followed their example. Helmets of leather and brass were adopted by cavalry units of many armies, although it is likely that the decorative aspect was thought to be more important than the defensive. Body armour soon followed suit, and most British cavalry had abandoned theirs by the end of the 17th century, although some units continued with theirs until the 18th century. The cuirass was re-introduced in 1821 for the Household Cavalry, although in Europe it had been retained by many countries, and units of the French Army rode into the First World War complete with cuirass and helmet.

The First World War saw the development of special types of body armour, with that of the Germans perhaps best known, for this was issued to many of the troops for use by snipers, sentries and machine-gun crews. It consisted of a large breastplate with several plates suspended from it, and weighed around 19–24 lbs. Many other countries also experimented with body armour during the war, but apparently never used it on the same scale as the Germans.

Collectors of military firearms are likely to find that it is not at all easy to locate specimens, for the demand has increased enormously and prices have rocketed. Swords and bayonets are still to be found at reasonable prices, although some rare examples naturally command a high price. For suggestions as to care, display and acquisition the reader is referred to *Small Arms* and *Swords and Daggers*, details of which are given in the bibliography.

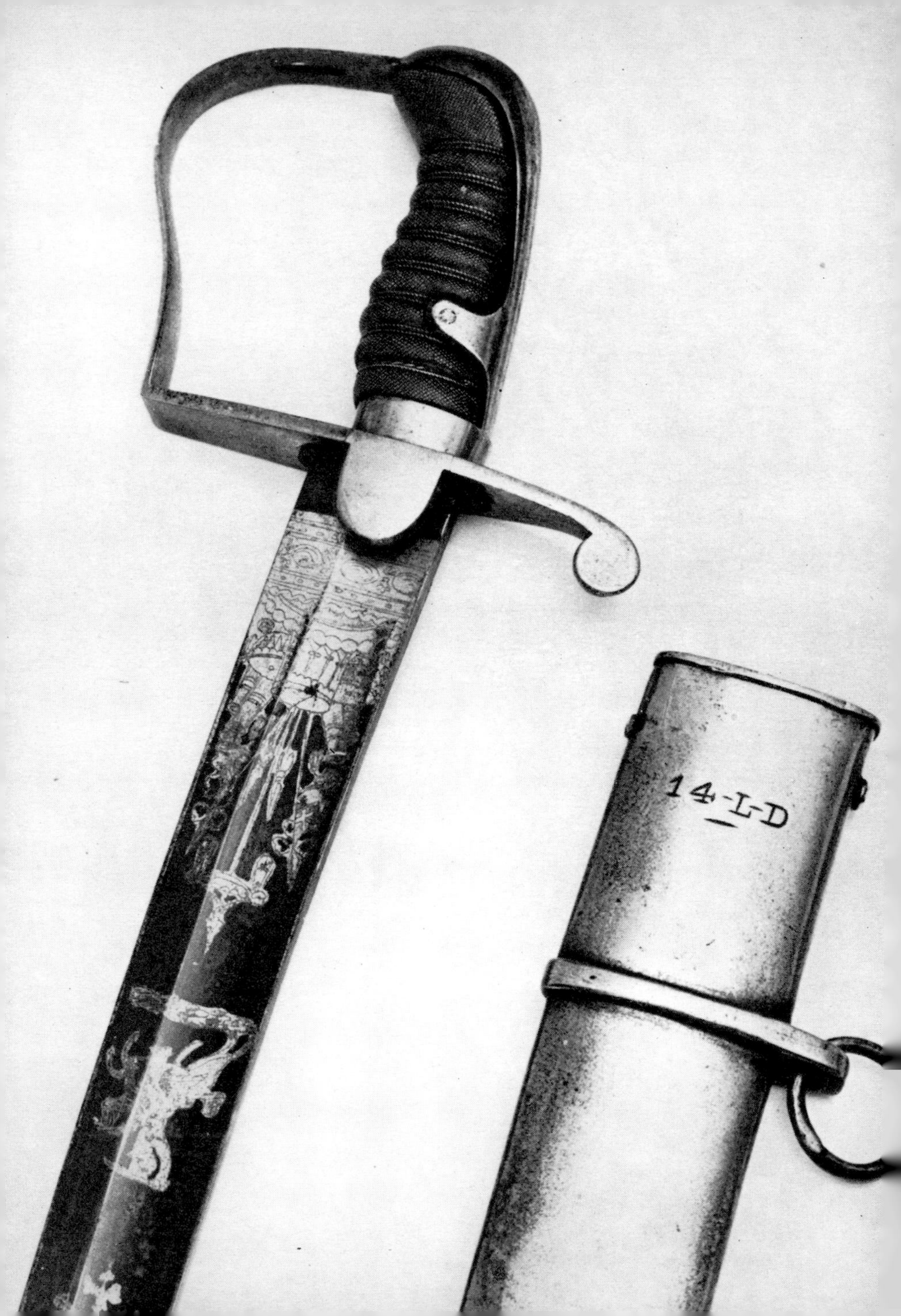
14-L-D

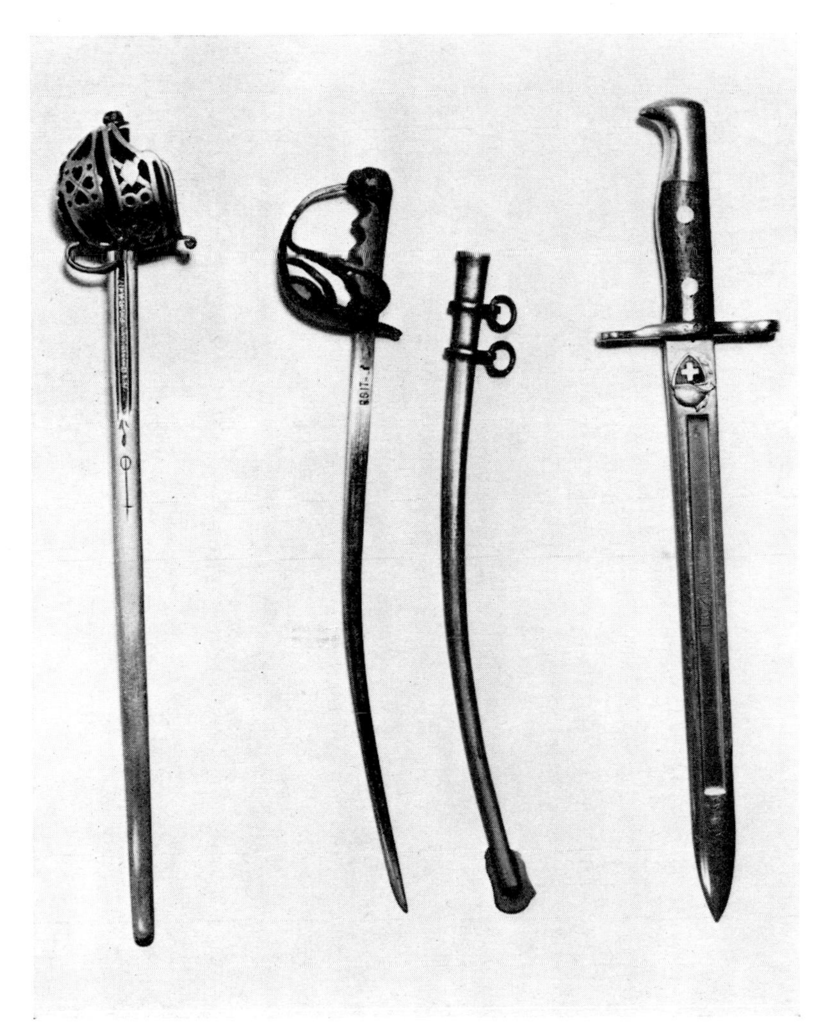

Miniature weapons: *Left:* a Scottish broadsword. *Centre:* Italian model of cavalry sabre and scabbard. *Right:* Letter opener in the form of a bayonet, made in Switzerland and bearing the date 1939.

◄

Stirrup hilted, curved steel sabre of the type issued to the British Light Cavalry in 1796. This weapon is unusual in that it is decorated with military motifs done in blue and gilt, which suggest that it belonged to an officer. On the steel scabbard is engraved 14-L-D, signifying the regiment, 14th Light Dragoons.

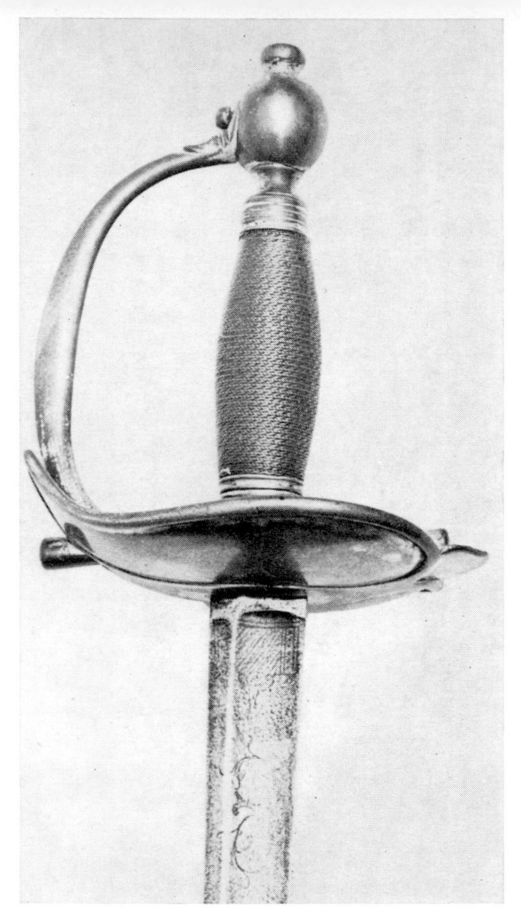

A "boatshell" hilt of gilt brass with a silver wire binding on the grip; dress sword worn by officers of Heavy Cavalry. The maker's name, J. J. Runkel, is engraved just below the hilt.

Studs and the monogram L.G. characterise this style of sword of the First Life Guards, used as early as 1834. This specimen is Victorian; the scabbard is of steel and has two lockets.

Brass hilt on a straight-bladed French sword intended for the Cavalry of the Line, made within a year of their defeat at Waterloo.

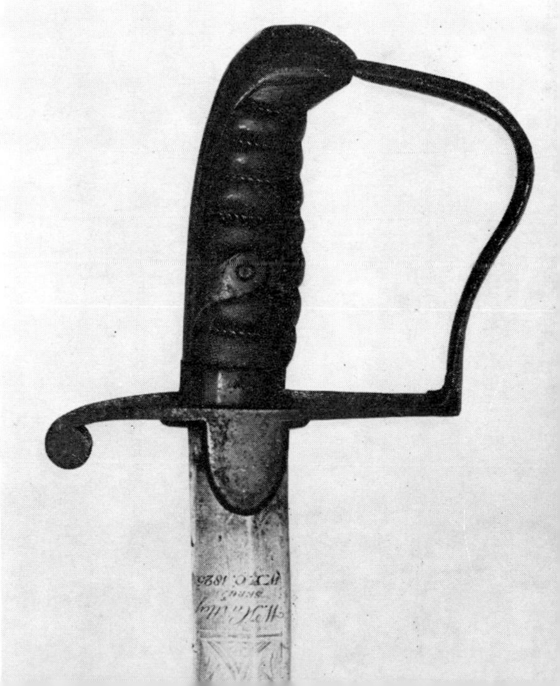

nother 1796 pattern Light Cavalry Sabre (see page 134), but used this time by a member of a local yeomanry unit after the style had en replaced in the Regular Army in 1821. Owner's name and regiment are inscribed on the blade.

Light Dragoon flintlock pistol made by Jordan and dated 1759.

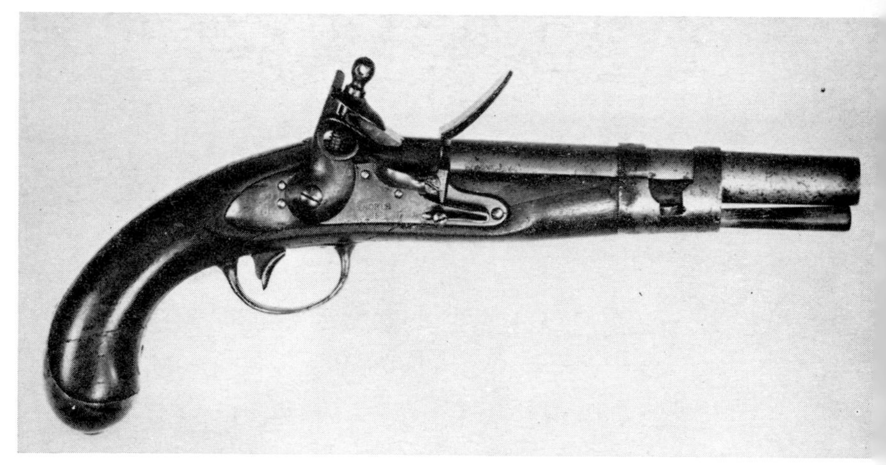

U.S. military flintlock, Model 1813, calibre 0·73 in., by Simeon North.

U.S. military pistol, Model 1808, Navy pattern, again by North; less French and more English in style, however, than the above.

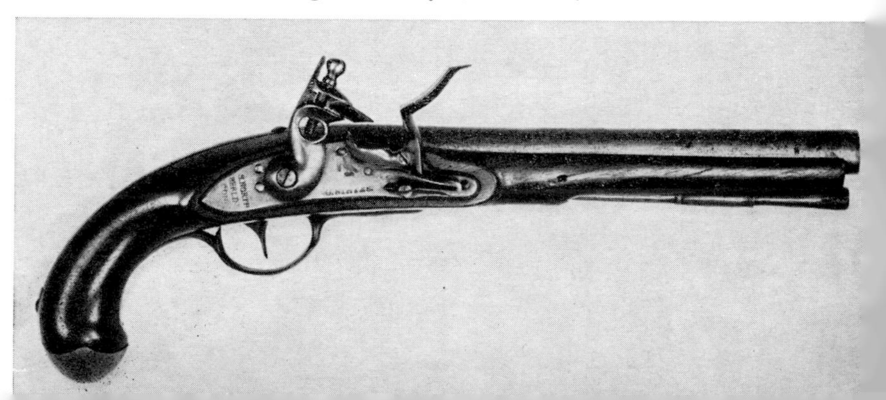

A rare British Military flintlock pistol bearing the cypher A.R. (Anna Regina) on the butt cap and lockplate. The maker's name, T. Fort, is engraved on the tail of the lockplate. Date is 1706.

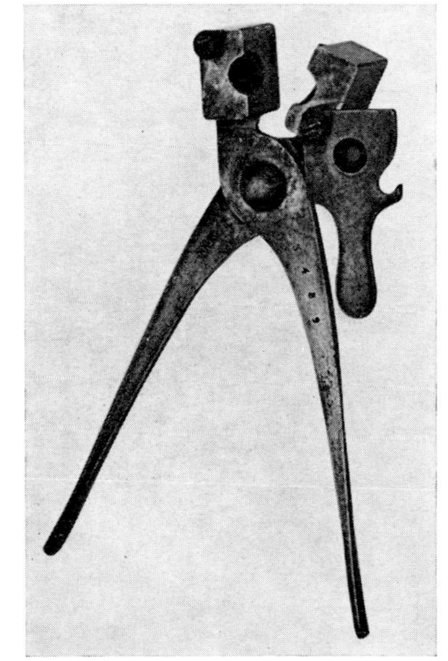

Bullet mould, mid-19th century, for casting lead bullets, calibre 0·577 in., to be used in the percussion Enfield rifle.

Russian sword with brass mounts for the hilt; black leather scabbard. The Russian swords were worn in a reverse style to that of the rest of Europe, so that the knuckle bow pointed to the rear rather than to the front. Overall length 39½ in. *C.* 1910.

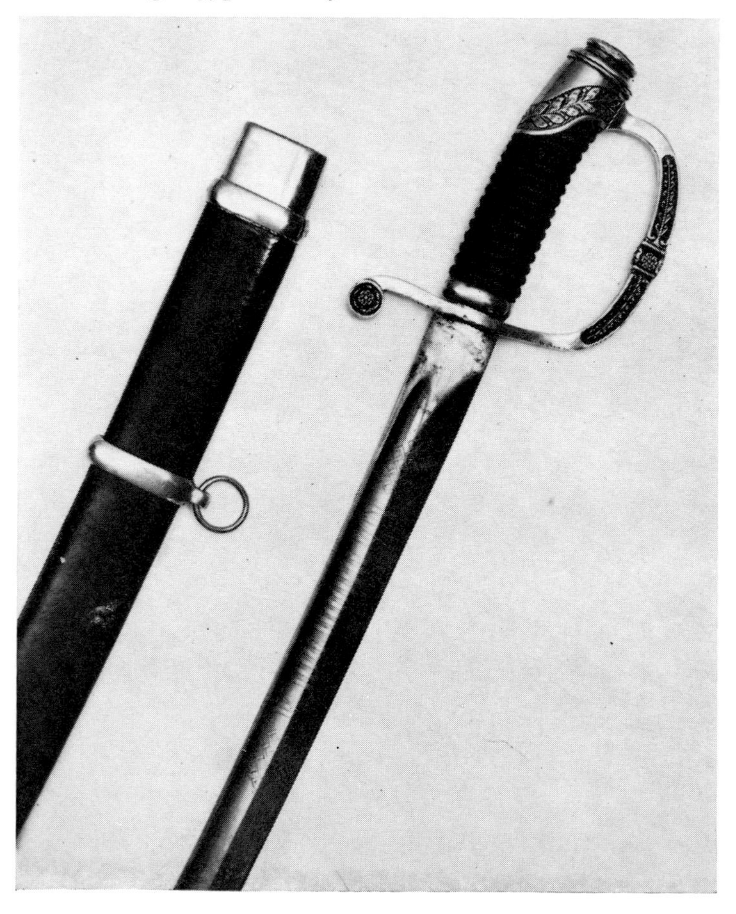

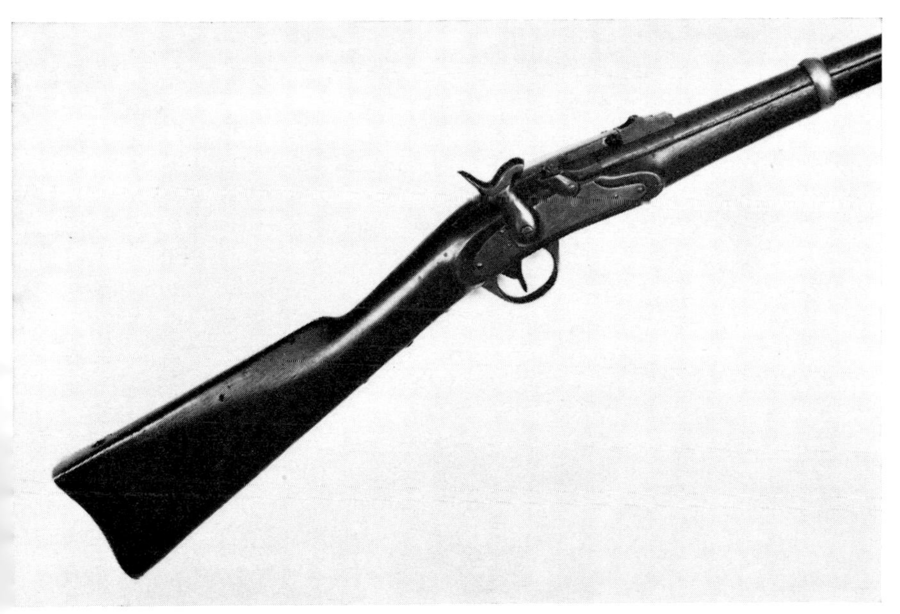

Butt and lock of a Merrill Carbine – a popular weapon during the American Civil War. The breech on this type of gun was opened by means of a bar which was situated above the lock.

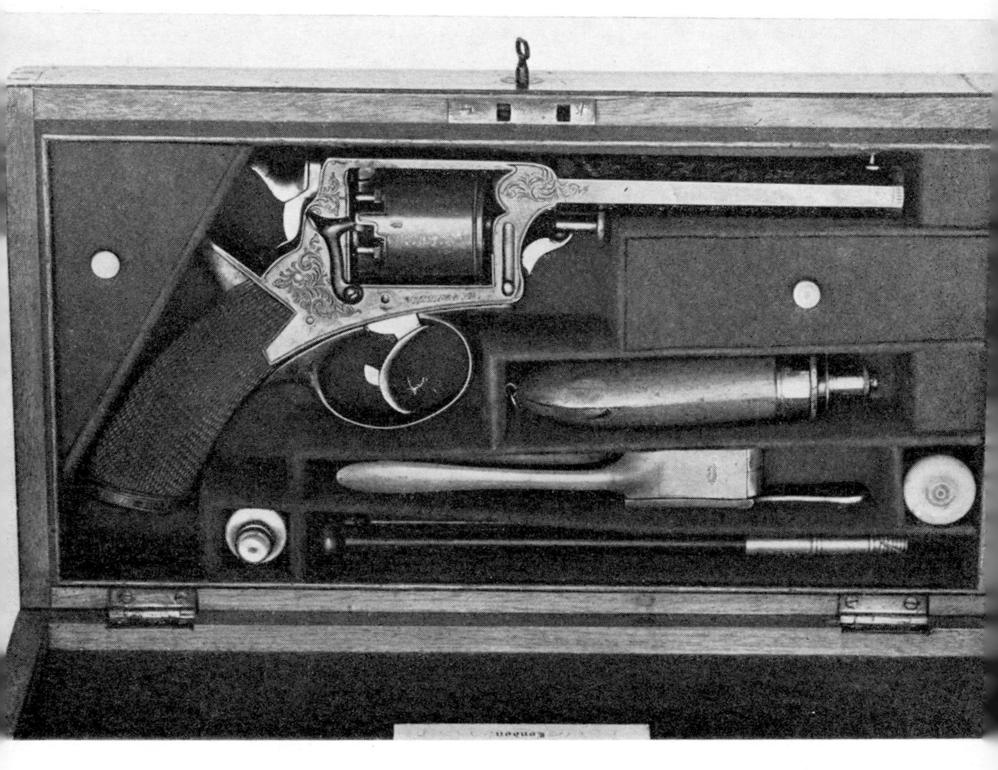

Percussion model revolver, patented by Tranter. In the case is a double bullet mould, powder flask, metal oil bottle, cleaning rod and cylindrical box (*bottom right*) for spare nipples or caps. *Below:* A close-up of the barrel, on which is engraved the name of the owner and his regiment: *W. A. Wynter 33rd Regmt. C.* 1870.

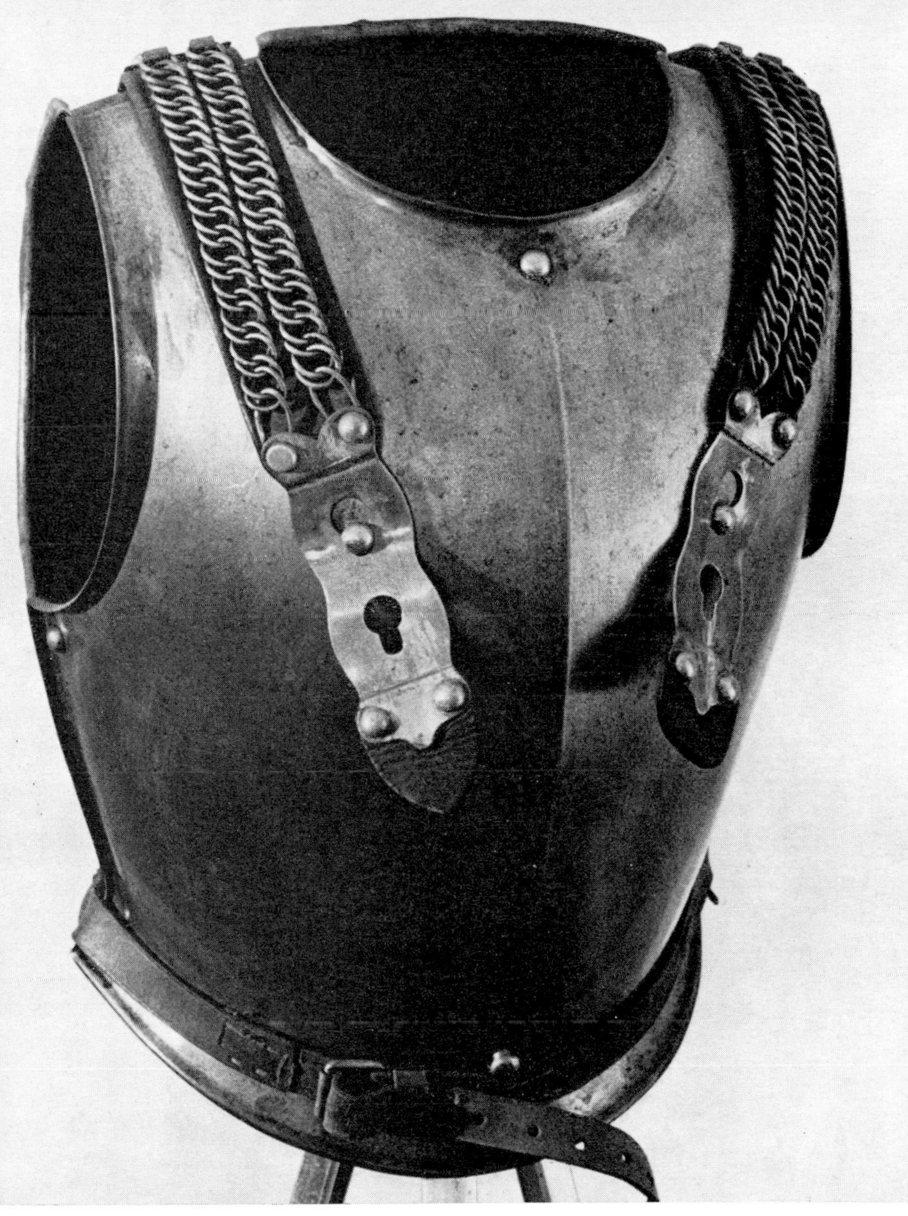

French cuirass, model 1854. In the British Army the cuirass was abandoned during the 18th century, but on the Continent it persisted until the First World War. There is very little difference between this, a Napoleonic model, and one worn during the First World War.

Dutch bayonet for the Mannlicher rifle. The leather scabbard and the rather narrow blade are, in fact, unusual features for this type of weapon. The bayonet is retained in the scabbard by a leather strap held in place by two loops.

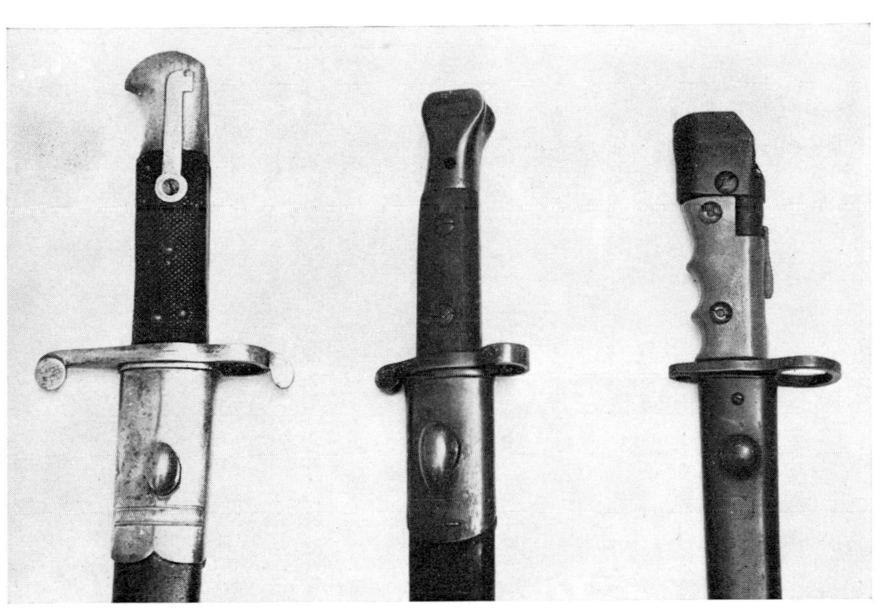

Three bayonets, *c.* 1860, 1890 and 1946 respectively, showing the gradual simplification and shortening of bayonets over the last century.

Modern replica of a bowie knife with inscribed blade. This type of weapon was commonly used by both sides during the American Civil War.

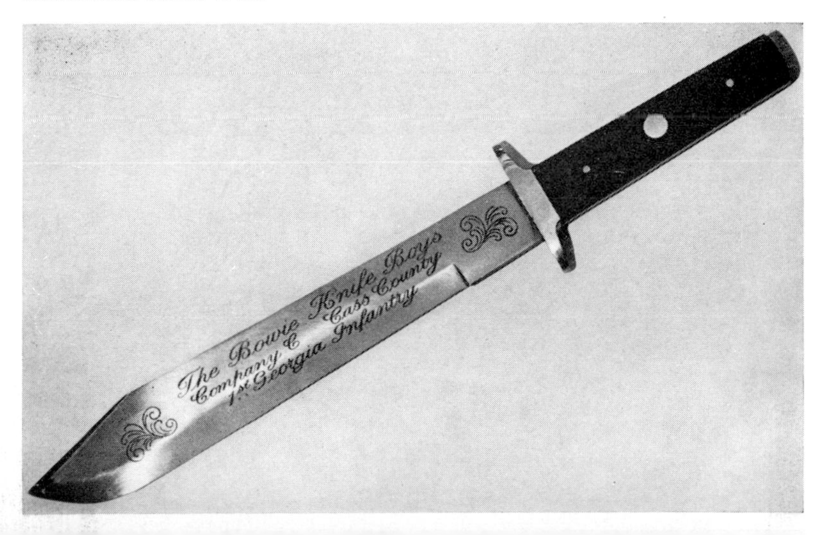

Dress helmet of the Gentlemen at Arms. The general shape of the helmet resembles that of the Cavalry Helmet of the Albert pattern.

VII

HELMETS AND HEADDRESSES

Helmets of one kind or another have been worn since earliest times, and appear on carvings, statues and wall paintings of the Egyptians, Sumerians and Assyrians. Leather was almost certainly the first material to be used, but as man's skill in metalwork developed he used these talents to strengthen the helmets. They were always something more than purely defensive, for they provided a convenient foundation for decoration; horns, feathers, spikes, fur and metal shapes have been added to helmets all over the world by most races. Some of these extras were purely decorative, but others were far more functional. Roman officers often had horsehair crests fitted on their helmets as a means of identification in battle. Much later this use of crests was to become one branch of the science of heraldry.

Starting at the Norman invasion, it is possible to make a consecutive study of helmets, for there are several fairly clear-cut and distinctive lines of development. Changes in tactics and alterations in weapon design all made demands on the skill of the armourer. It was realised that thickness of metal was not the only means of supplying protection – shape and design were very important factors, for a smooth, curving surface could turn the edge of a sword and absorb, or deflect, the greater part of the impact. Greater protection for the face was provided, and the single bar, the nasal, of the Norman helmet was widened, as were the neck and cheek pieces, until all united to form a helmet which completely encircled the head.

Helmets of this basic style were produced in a variety of shapes, but all had slits or holes for vision and ventilation and had some form of internal padding. These early helmets were large, heavy, hot and uncomfortable to wear, so that they were worn only when it became necessary, and were normally carried by the squire or on the saddle. One of these helmets was the great helm, of enormous cylindrical shape, that completely covered the head and reached down to rest on the shoulders. Great helms were frequently secured against accidental loss in battle by a length of chain fastening them to the breastplate. Great helms afforded a great deal of protection, but were so cumbersome that they were soon relegated to the sporting tilt and simpler, but equally efficient forms designed for war.

During the 15th century a particularly graceful and attractive style of armour, usually described as Gothic, was in vogue, and with it went that type of helmet called a sallet. This name covers a number of variations on a basic shape: some are simple, others quite complex, with gracefully laminated tails. Contemporary with the sallet were the armet and close helm, both of which envelop the head. Close helmets survived until well into the 17th century, although the later ones tend to be coarser in shape. Ingeniously designed close helmets could be hinged open to facilitate their donning or removal.

It is with the appearance of the close helmet that the collector can begin to take an active interest, for these helmets, although expensive, do turn up in the sales rooms and antique shops. Obviously the earlier ones are extremely rare and expensive, but those from the end of the 16th century are less rare. Even so, in view of the probable price, it is always as well to check them very thoroughly and, if necessary, seek a second, expert opinion before making a purchase.

Close helmets were part of the equipment of the fully armoured knight, but the increasing use of firearms was changing the style of warfare, and the heavy, rather slow knight was becoming less and less important. The last armoured knights to be seen in any numbers on the battlefields of Western Europe were the cuirassiers of the wars of the first half of the 17th century. These troops wore three-quarter armour reaching to the knees and some form of close helmet. Their cuirass, breast-and back-plate, and helmets were fairly heavy, for they were thick enough to withstand a pistol ball fired at them. Cuirassiers were gradually replaced by faster, light cavalry, equipped only with cuirass and a light, open-faced helmet called a burgonet. Such helmets were very popular, and the style known as Zischägge, or lobster pots, was common wear on both sides in the English Civil War.

Infantry helmets were generally simpler than those of the cavalry. Most foot soldiers wore a morion with its cone-shaped crown, and a brim of various width, often with cheek pieces attached. These "pots" were produced in quantity for the pikemen of the period, and were frequently of rather mediocre quality. One useful point in dating these helmets is the method of construction, for earlier examples were fashioned from one piece of metal, while later ones are made from two separate pieces of metal riveted together.

In England, the infantryman had generally discarded his metal helmet and breastplate by the 1670s or 1680s, although it appears to have been a somewhat haphazard change, and no specific order for discarding them is known. It is about this same period that the Grenadiers of each company – picked men specialising in throwing the small hand bomb, Grenado – began wearing a furred cap. This cap is mentioned as early as 1678, but, as is so often the case, the recorded description is imprecise. It may be safely assumed that it was something like the later busby with a cloth bag issuing from the centre of a fur cap. Later the cloth bag was stiffened to make it stand up in a cone shape, and in this style lies the beginning of the famous Grenadier's mitre cap. Apart from this rather special case, most of the other infantrymen wore a hat that was more or less civilian in style, although most seem to have had some decoration in the form of lace or ribbon.

Cavalry headdress was essentially the same as that of the infantry, although the civilian style cocked hat was preferred. Again the hats were frequently embellished with lace, ribbons or buttons, and, in the case of officers, often with a feather. Many of the cavalrymen wore beneath their hats a metal skull cap, a secrete, to provide greater protection for the head.

These two basic styles of the Grenadier's mitre cap and the tricorne continued in general use for most line regiments throughout the first half of the 18th century, although some distinction between regiments was made. Embroidered patterns on the Grenadiers' caps varied, and the number of the regiment was shown in either Arabic or Roman numerals. In 1751 a Clothing Warrant was issued, and in this it was stipulated that the front of the cap should be the same colour as the facing, or coloured material, on the tunics; it should have the King's cypher, the Hanoverian white horse, and the motto "Nec Aspera Terrent". Eight different colours of facing are listed, with yellow as the most popular, being used by eighteen regiments; next comes green used by nine regiments. The same type of mitre cap was worn by the Horse Grenadiers, but most of the cavalry still wore the ordinary three-cornered hat. One distinction was made in the case of the Light Dragoons raised in 1756, for they were to wear the so-called jockey cap. This was to have the royal cypher in front in brass, as well as the number or rank of the regiment. It was also to have a tuft of horsehair, half red and the other half of the same colour as the facings of the regiment. The skull was rounded and fairly close-fitting to the head, and this style was later used by a number of other countries.

These basic styles of headdress were common to many of the armies of Europe, although, naturally, there were numerous fine distinctions, and in the styles of the 1770s can be seen the first signs of the profusion of styles that was soon to burst upon the military scene. The old tricorne hat was changing to the cocked hat, with the wide brim folded up at the back and front, and worn with the points to the side or running back to front. This was the style that has survived in a modified form with many of the diplomatic and naval dress uniforms of today.

In North America, until the War of Independence, the main armed forces were, of course, British, but following the signing of peace, Congress was able to turn its attention to the matter of military uniforms for the United States. The style of uniform was not dissimilar to that of the British forces. Cocked hats, jockey caps and tricornes were still general wear. However, there was one major change taking place about this time, for in both the American and British Armies there was an increasing use of trimmings on hats; feathers, tufts and lace were all adopted in a variety of patterns. It was finally decided by the British War Office that some standardisation would have to be enforced, and in 1796 the lace was removed from the British infantry hats. In 1797 it was stated that in future Battalion Officers were to wear red and white feathers, Grenadiers white and Light Infantry green. It was during the Napoleonic Wars that styles proliferated, and the result was a bewildering variety of designs. In Britain the situation was complicated even further by the large number of volunteer groups.

In 1793 France had declared war on Britain, who formed a coalition of nations to crush this new republic. Fighting, predominantly naval for Britain, continued until March 1802, when the Peace of Amiens was signed. This peace was short lived, and hostilities were resumed in May 1803. It then looked as if the French would certainly invade Britain, and there was an immediate national reaction as, all over the country, groups of local gentry formed enthusiastic volunteer groups. The same response had been evoked by a similar threat to Ireland in 1778, and volunteers had been formed, although the original spirit changed, and by 1793 they had become mainly political in nature and were eventually suppressed by the Government. British volunteer groups were originally authorised under the Act of 1794 (34 Geo. III c. 31), which set out conditions of service and supply of materials. With the Peace of Amiens, the Acts governing their service were automatically ended, for they had been only for the duration of the war. However, in June 1802 another Act (42 Geo. III c. 66) was introduced, and this stated that in future volunteers would serve without the meagre allowances that they had previously received. This position was modified in March 1803, and in general it was agreed that the Government would supply arms and accoutrements – in fact, many groups purchased their own – but the volunteers would be responsible for their own uniforms. As each new group was formed, an enthusiastic committee was set up to discuss and design their uniforms and helmets. The resulting products were generally rather gorgeous and frequently impractical, with plumes, feathers, lace and a galaxy of colours. With the final victory at Waterloo in 1815, the obvious need for such groups was ended, and, in fact, the volunteers were stood down in 1814, to be followed in 1816 by the Local Militia, a more regular body.

For over twenty years the country had been at war, and by now the tradition of fancy headdresses had become well established. Bearskins, 20 inches deep in the case of the Life Guards, were very popular from the end of the 18th century onwards, and continued to be worn by some regiments right up to the present. The early bearskins, worn also by some of Napoleon's troops, were fitted with a large gilt plate embossed with arms or regimental badges. Around the top, a plaited cord and tassel was gracefully draped. Metal helmets were commonly worn by many cavalry regiments, and these were of white metal or gilt with added decoration in the form of laurel leaves, lions' heads, horsehair plumes, a great comb of bearskin, tassels and tufts. Dragoon regiments were also issued with helmets in about 1822/3. Hussar regiments, based on the dashing Hungarian light horsemen, were formed from some of the Light Dragoons, and these wore the busby with its fur muff and coloured cloth bag, which was originally free-hanging. In later styles, the bag was secured to one side of the fur muff. For the officers, there was a tall feather, and for the men, small tufts in white or red.

At the beginning of the 19th century the exigencies of combat soon forced the authorities to realise that some more practical form of headdress was required, and about 1800 the shako was introduced for the infantry. This was a tall leather stovepipe-like hat, with a small peak. Some twelve years later the height of the shako was reduced, but a false front was fitted to give the appearance of height. Most of the shakos had a draped cord, white for most of the infantry, crimson and gold for the officers and green for the light infantry. In 1816 the bell-topped shako appeared, with its rather impractical wide crown. Most of the shakos were fitted with some form of plume, but these were replaced in 1835 by smaller coloured pompoms.

Shakos remained general wear, although the style was varied considerably, and the Albert shako reverted almost to the original pattern. In 1855 the height of the shako was reduced, and instead of its usual tubular shape the back was tapered towards the top. In 1861 this shape was retained, but the height was reduced even further. Eight years later this form of shako was modified, and braid was added around the top and down the sides – red and black for men and gold for officers.

Fashion in military matters tends to be influenced by the nation which is militarily predominant at that particular period, and in the 1870s the Prussian army held pride of place after its spectacular defeat of the French. This may well explain why in 1878 the British infantry were issued with a spiked helmet. This was of cork, and had a peak at the back and front; that in front bordered with a narrow brass edge. For most regiments the helmet was covered with dark blue material. In the case of the light infantry the material was dark green, and the rifle regiments had green with bronze fittings. The Home-Pattern helmet was fitted with a curb chain or chin strap fitted with links $\frac{5}{8}$ inch in diameter, and this could be worn under the chin or across the front of the helmet and secured to a hook behind the spike. All the illustrations show the strap as passing straight across the front of the helmet; it is a strange fact that the majority of those surviving have a chain which is too short and when hooked up crosses over the badge.

Distinction was made in 1881 when the Artillery were given permission to replace the top spike by a ball and cup fitting. The spiked helmet was replaced in different regiments at different dates, lingering on with some until the 1920s, when the cloth cap which had first appeared in 1902 in the form of the Broderick cap became general wear. The earlier style was very Germanic and lacked a peak, but by 1905 the peaked general service cap was common. Tanks with their limited cabin space led to the introduction of the beret, and this became more or less general issue from 1943.

The First World War spelled the end for most of the glamour of military headdress. Gone were the busbies of the Hussars and the metal helmets of the Dragoons with their variety of coloured plumes.

The development of British headdresses was largely mirrored by the United States, although the bitter experience of the American Civil War brought home to the forces the importance of practicality. Many of the early flamboyant styles affected by some units of the Union and Confederate forces soon gave way to the simpler cloth kepi and slouch hats.

A brief survey such as this cannot even attempt to cover the myriad of regimental distinctions, and the best sources of information are the various dress regulations. Unfortunately these are not always to hand, and it may be necessary to rely on secondary sources, such as the books listed in the bibliography.

Generally speaking, a collector may hope to acquire specimens which date from the 19th century at a reasonable figure. During the 1850s and 1860s there was a resurgence of the volunteer movement, and samples of the equipment of such bodies are not uncommon. Those of the regular regiments are perhaps a little less easy to acquire. There is, of course, a growing tendency for prices to rise as demand and interest grow, but, in general, it is better to acquire a good specimen, complete with all accessories, even if it means paying slightly more. Given time, it may well be possible to locate some missing item, but, in fact, it is usually rather difficult and frustrating. In any case, single items such as plumes are frequently quite expensive. Officers' headdresses still turn up complete with their japanned tin boxes, and these are convenient for storing and protection.

Display is very much a matter of personal taste; some collectors like to display the helmet on its own, others prefer to mount it in some way. Some use no more than a wire frame, but others like to create more substantial heads fashioned in a variety of media, such as paper sculpture, expanded polystyrene, papier mâché or plaster. Cleaning is very much a matter of common sense – often a good brush is all that is required. Some specimens may need gentle washing, using a stiff brush and warm water. In all cases, care is essential. Moths are an obvious danger to the cloth helmets, but modern deterrents may be used with confidence.

Leather becomes very brittle with age, and should be treated with neatsfoot oil or, better still, one of the proprietary leather dressings used by the museums services. Brass and silver fittings and badges may be cleaned with one of the long-lasting polishes, but it must be remembered that every cleaning represents a slight wearing down of the surface, and it may well be necessary to consider whether a good-quality clear lacquer is not preferable to continued cleaning of metal parts. Dust and atmospheric pollution can affect both material and fittings, and if it is impossible or undesirable to display the helmets under glass, then a clear plastic bag may prove a useful compromise.

Reference books are an absolute necessity for the collector, for many helmets and headdresses have, over the years, been used for theatrical purposes, and may well have been altered or adapted. It is important to be able to recognise the component parts, or indeed identify any helmet, and for this good books are essential.

Value is an extremely difficult thing to assess, but the three governing factors are rarity, condition and demand. Genuine examples of early military headdresses are rare, and consequently expensive, whereas cloth helmets, especially those of the volunteers and militia, are still comparatively inexpensive. Continental examples, with the exception of French and German, are less valued and consequently cheaper, although items such as Imperial and Nazi German helmets and headdresses are in particular demand and their value is steadily increasing. However, it is important to be able to identify the great variety of models, for some, although superficially similar, are so much rarer than others. Metal helmets are very much more expensive, and all too often lack component parts, such as chin scales or plume holders.

A headdress or helmet possesses a unity of its own; it is complete in itself. This feature, together with the decorative qualities, make helmets extremely attractive and satisfying items to collect.

Fine example of a Cuirassier's armour. It is of heavy construction; the breastplate alone weighs over eleven and a quarter pounds and bears a dent made by the bullet that was used to "prove" the metal, or demonstrate its strength. The skull is made of two pieces united at the comb, rather than being forged from one piece of metal, as was done earlier. This specimen is from Germany, *c.* 1620.

Sallet, German 1450–60; a graceful helmet forged completely from a single piece of metal. This specimen is without a separate visor.

◀

Superbly decorated example of a gorget, a plate of armour made to guard the neck and throat. It is embossed with detailed battle scenes depicting an assault on a town, further embellished with gold and silver and made in Milan early in the 17th century. The wearing of this plate, often in a decorative form, lingered on after most armour had been discarded. The lower part, which covered the neck and top of the chest, was in use until the 19th century.

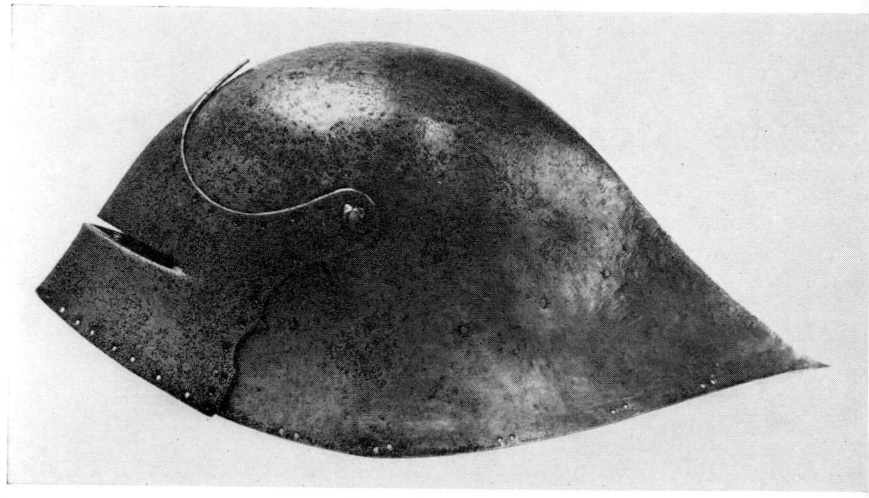

Sallet with sweeping shape designed to permit it to be pushed to the back of the head to avoid the necessity of lifting the visor. A series of holes around the edge originally held rivets for securing the inner lining and straps. The face guard, or visor, is pivoted so that it could be raised or locked down in position by means of a spring clip. This specimen is German, 1480–1510.

▶

Brass helmet with silvered badge, crest and chin strap worn by the French mounted Gendarmerie about 1910. In style it is basically the same as that worn by many French cavalry units right up to the First World War, though they usually had a dangling horse-hair crest.

Pikeman's helmet, or pot, of such quality to suggest that it was worn by an officer, for most examples are without fluting. At the back is fitted a holder for a plume. French or Italian, *c.* 1630.

◀

A morion, one of a group of five that belonged to the Trabanten Guard of the Electors of Saxony, *c.* 1580. This type of helmet with a high central comb was commonly worn by foot soldiers, but this example is of superb quality, and the surface is gilded and etched with bands and panels and bears the Arms of Saxony.

▶ French helmet with side plume of the Sapeur-Pompier or Fire Brigade, which in France is counted as part of the armed services. It is, in fact, very similar to the helmet worn by the French cavalry.

A French helmet, *c.* 1880, similar to the British kepi, or shako. The fittings are of gilt; the cap top is of silver facing.

QUO FAS ET GLORIA

▲

Officer's forage cap of blue cloth with gold embroidery on the peak. The badge is of the Norfolk Regiment, the old 9th Regiment of Foot. This example conforms to the dress regulations of 1883, and from the evidence of the maker's label, dates from before 1890.

◀

Blue cloth helmet, home pattern, as worn by the Royal Artillery, Field or Garrison, and dating from 1901. The curb chain may not be the original, for it should sit straight across the helmet's peak.

Shako of the Highland Light Infantry, *c.* 1900. The badge bears the title ASSAYE over an elephant, and was granted to the regiment in honour of its service at the Battle of Assaye in 1803.

▶

Fine example of an officer's helmet of the Inns of Court Regiment. Here the chin strap is the original and sits comfortably across the peak under the side hook. The style of crown dates this 1878–1901.

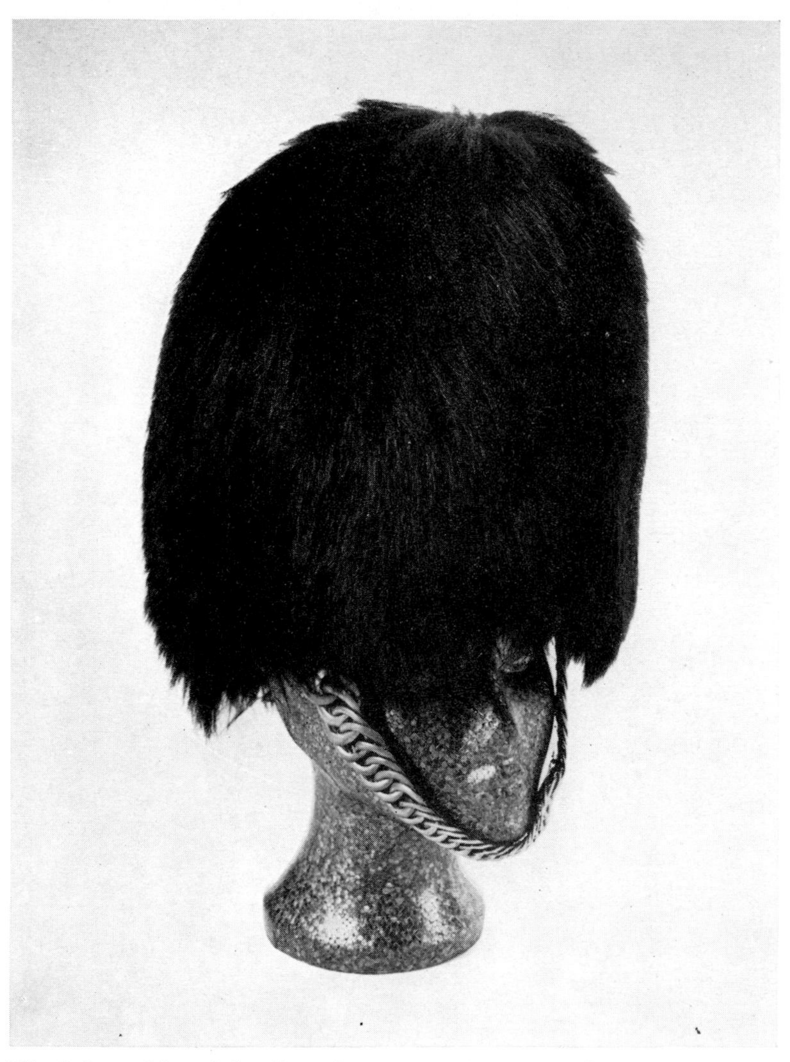

Black bearskin of the Guards – a popular style of headdress, having been worn by regiments in most European armies at some time or other.

Cap of N.C.O. of 12th Lancers; dated 1902. These ornate headdresses were of black leather; red cloth covers the top half of this example. The cap lines at left pass round the body loop on the left breast.

Helmet of the 7th Dragoon Guards of the style worn at the time of the Crimean War. In gilt brass with a diamond cut star on the front, it has a black-and-white plume, the colour worn by the regiment.

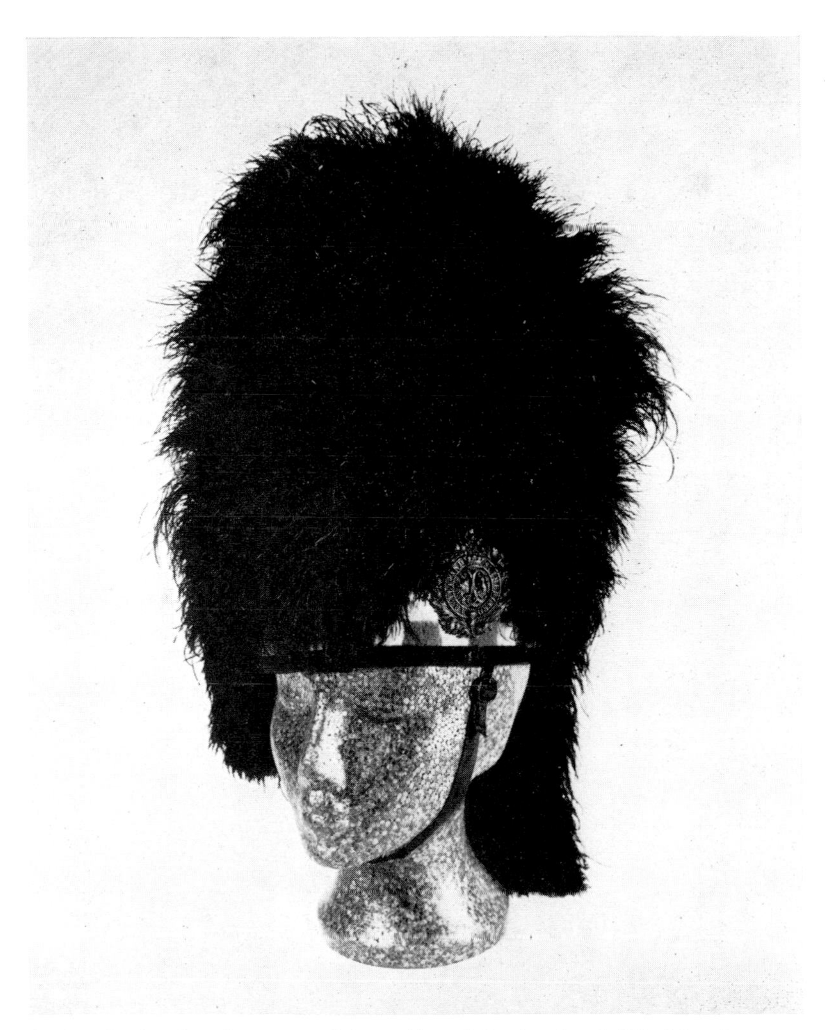

Officer's feather bonnet of Argyll and Sutherland Highlanders, *c.* 1900. The bonnet was built up of ostrich feathers, and on the left side was the regimental badge and a hackle feather 8 in. long.

German *pickelhaube*, 1914–18; leather with the usual spike. The metal scales on this example indicate that it is an officer's pattern.

German steel helmet, First World War. Introduced as a result trench warfare, it was soon adopted by most of the belligeren nations.

French steel helmet, virtually identical in both World Wars, but less sturdy and more complex in construction than other patterns.

ɔi, or shako, of the Queen's Westminster Rifle Volunteers complete h the pompom that replaced the plumes worn previously.

Japanese steel helmet, Second World War. The anchor stencilled in yellow paint suggests that this helmet was for marine or naval use.

Facsimile of a Confederate forces' kepi, American Civil War 1861–65. The badge is upside down, as the hilts should be next to the peak.

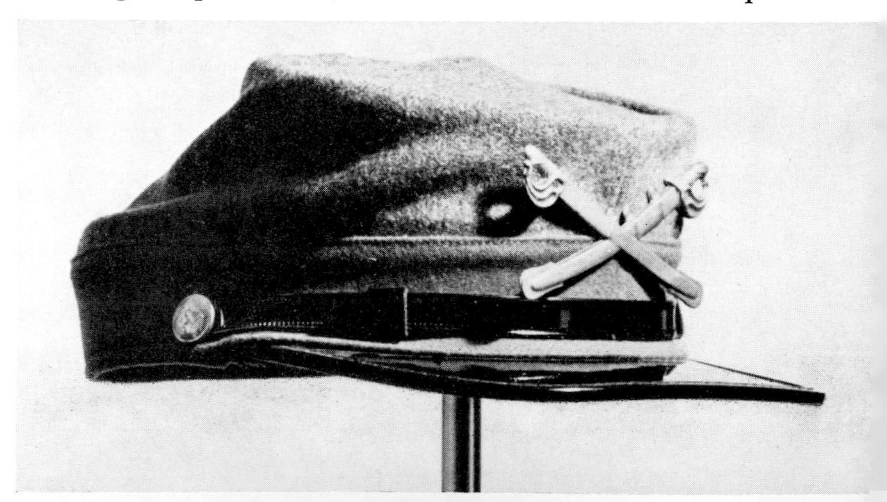

Plum-coloured kepi, Mexican, *c.* 1900, with embroidered oak leaves and an eagle carrying a snake in its beak at the front.

Facsimile of a Union forces' kepi, American Civil War 1861–65. Blue cloth is the main material; the chin strap and peak are of leather.

◀

Many military academies still retain, for ceremonial purposes, a picturesque style of uniform. This kepi, from the uniform worn at an academy at Saint Cyr, in France, is basically the same as that worn by many armies during the middle of the 19th century.

Officer's uniform; 7th Light Dragoons, *c.* 1786.

VIII

BELTS, BADGES AND BUTTONS

From the very beginning soldiers have had to carry various bits of equipment, and they have all used some form of belt for their swords, quivers or daggers. Ancient civilisations used both waist belts and shoulder belts, but medieval sword belts were commonly worn around the waist. They were of sturdy construction, frequently complicated in design and elaborate in decoration. By the late 16th century the lighter rapier was replacing the heavier cutting or slashing sword, and the older, heavy belt was no longer required. Now a narrow girdle encircled the waist and was fitted with a metal loop to which was attached a wide, multi-strap hanger holding the sheath. By the middle of the 17th century a simpler version, with a double strap hanger, was in use. Many officers preferred a wide sling which crossed the body from the right shoulder to the left hip – this, too, ended in a double strap to which the scabbard was fastened by a stud or hook. These slings were frequently decorated with lace, velvet or coloured leather.

Officers and most cavalrymen retained their shoulder belts, but the infantry were issued with a waist belt worn outside their overcoats and from which, at the front, hung their plug bayonet in a small leather sheath. The sword was also attached to this leather belt by two short extension straps. Over his shoulder the infantryman also wore a wide cross belt from which hung his bullet pouch and cartridge bag – a natural development from the bandolier carried by the early musketeers.

This basic pattern remained throughout much of the 18th century, although the cavalry frequently carried a second shoulder strap fitted with a spring clasp, and from this hung their short musket or carbine. Early in the 18th century many of the infantry began to wear two cross belts – from that on the right hung a pouch, while the left carried a double sling holding the bayonet above and a sword below. By the 1770s most of the British infantry no longer carried a sword, and the two belts now supported only the bayonet and pouch. However, the practice was not uniform throughout the Army, and there were numerous variations; in general, it may be said that officers and sergeants had only one cross belt, while privates had two. Regulations for the Accoutrements of the Regiment of Foot Guards and Marching Regiments, 1784, specified that the cross belts were to be of buff leather, 2 inches wide, and that part of the belt in which the bayonet was carried was to be made to slip off and on with two loops. The cartridge box was to be "by way of a magazine" and to be fixed to the bayonet belt "easy to put on", and was to be worn only on the march and on active service.

Around the middle of the 18th century there first appeared a decorative plate, known as the shoulder-belt plate, worn at the centre of the cross belts on the chest. Most of the plates bore the initials, number or badge of the regiment and, in many cases, all three. Some have these details engraved on them, while others had the badge made as a separate piece and secured to the flat plate by a split pin (*at the back*). The use of these shoulder-belt plates was nearly always restricted to infantry regiments, and design was usually peculiar to a regiment, sometimes to each battalion. Most were made of metal whose colour matched the gold or silver lace worn by the regiment. Volunteer units also wore these shoulder-belt plates, but in most cases they seem to have been a little more elaborate and decorative than those of the regular forces. Originally the usual shape had been oval, but by the 1820s most were square. Shoulder-belt plates were finally abolished in 1855.

During the 19th century the pattern of waist belts became standardised, and details were set down in Dress Regulations. Most had a round tongue and buckle type of clasp, usually bearing the name of the regiment. Some were of patent leather, while others, worn for dress occasions, were some 2 inches wide with a set pattern of coloured lace on the leather. Although shoulder belts for swords were discarded, many regiments continued to wear a pouch belt which crossed the shoulder and had a pouch which fitted into the small of the back. Such pouches bore the regimental badge fitted to the flap. Many units, in particular rifle units, had a small whistle, secured by a length of chain, fastened to these pouch belts at the front, and this fixture was particularly popular with the Victorian volunteers.

In the early 1860s Sir Samuel Browne – he was awarded the Victoria Cross during the Indian Mutiny – designed his famous belt which came into use around the 1870s and was adopted by most armies in the world. The Sam Browne belt was of stout leather worn originally with two cross braces over the shoulders, but at the end of the 19th century most regiments had discarded one of the braces except when wearing a holstered revolver. The Dress Regulations for the Indian Army 1918 describe the belt as having a waist belt $2\frac{1}{8}$ inches wide and a length to suit the wearer, double tongue brass buckle, four brass Ds for the shoulder belt, two at the back and one at each side, a running loop, brass loops for attachment to the frog, shoulder belts plain, crossed at the back through a loop and both just over 1 inch wide.

Cavalry regiments always regarded themselves as being superior to the infantry, and their uniforms reflected this belief. Cavalry uniforms were usually more elaborate, with extra fittings, and early in the 19th century yet another was added in the form of a sabretache. This large flat pouch was suspended from the waist belt by three long straps and had a flap which covered a series of deep pockets intended for notes and maps, but which became, in fact, just another pocket for a variety of personal items. The outside flap was embroidered with the royal cypher and crown, and in Victorian times the practice of adding battle honours was common. Regimental designations were also embroidered on, so that the sabretache became very ornate indeed. For undress purposes a plainer model of patent leather with metal badges was used. Sabretaches were finally abolished in 1901.

Until the mid-19th century personal holsters were seldom used, for it was usual to carry the pistol in two holsters secured in front of the saddle across the horse's neck. Holsters of the 17th and 18th centuries were generally elaborately embroidered and decorated with fringes. The advent of the percussion revolver in the mid-19th century produced a more personal defence weapon, and it became usual to carry a plain leather holster attached to the waist belt.

Belts as a whole are not popular collectors' items unless they have some special regimental significance. Sam Browne belts and leather holsters are of no interest unless there is some special association, but the dress belts, pouch belts and sabretaches are of some decorative value, and there is more interest in these. Early examples of holsters are ornate and attractive, but their high price reflects these qualities as well as their rarity.

Far more popular for the collector of limited means are badges, for they represent an extremely cheap and interesting field of opportunity. Many of the more modern badges are still readily available at very reasonable figures, and may even be purchased in bulk at auction sales. The development of badges and uniforms were very closely related. Knights wore an identifying device, the crest, on their helmets, and a similar device was often painted or embroidered upon their tunic. Later simpler devices, such as sashes and favours, were worn as marks of identification.

Embroidery was the medium used for the earliest badges worn

on the headdresses of British troops, for it was decreed that the mitre cap should carry the regimental number in 1751. In 1768 another warrant gave permission for the regimental farrier to wear a badge in the shape of a horseshoe on the front of his cap – possibly one of the earliest examples of trade badges in the British Army.

Metal badges for headdresses, helmet plates, began to make their appearance in the latter part of the 18th century, but it was the shako with its large metal plate that offered so much scope for displaying badges. Gradually the plate was modified and finally acquired the star-like shape popular in the 19th century. Most helmet plates were large, but with the introduction of the general service cap they were reduced to a more reasonable size, and these cap badges are probably the most collected of badges today. At present they are still fairly plentiful and cheap.

During the 17th and much of the 18th century colonels commanding regiments were pretty well supreme in equipping and organising their own regiments; thus the regiments were designated with such titles as Colonel Hodges' Regiment or Colonel Wynn's Regiment. Since most of these colonels were originally titled men, it was natural that they should use some part of their coat of arms as the identifying symbol for their regiment. Many of the present-day badges owe their beginnings to such coats of arms. In other cases the badge commemorated an event in the history of the regiment. Thus the badge of the Scots Greys bears an eagle above the word "Waterloo" in recognition of their great bravery at this battle, when they captured one of Napoleon's eagles. Yet another type of badge is that which symbolises the service offered by the regiment or department; the artillery badge naturally carries a cannon and the parachute troops, wings and a parachute. Smaller badges were worn on the collars of the tunics, and these were made in pairs with the figures facing inwards towards the opening at the throat. In addition to regimental badges there are a number of tradesman and proficiency badges, thus the crossed rifles is the badge of the marksman, crossed swords, originally of the sword instructor, now specify the physical training instructor. So far all badges discussed have been fairly substantial and made of metal. Brass was, of course, most popular in the early days, but later other metals were gradually introduced, until the final degradation for generations of old soldiers came with the issue of a plastic badge which required little or no cleaning.

During the First World War there began the practice of identifying units by cloth shoulder badges bearing a variety of designs and insignia. Metal letters had normally been worn on the shoulder straps, but these were gradually discontinued after the First World War, and by the Second World War cloth badges predominated. Apart from divisional signs, the names of regiments and units were also produced on embroidered shoulder flashes, and these are still worn on battledress today. Such shoulder flashes are very common, and may often be purchased in bulk at auction sales.

Badges of rank in the 19th century were of metal, worn on the epaulettes of the tunic. The Dress Regulations for 1883 lists them all from the field marshal with his crossed batons, the colonel with a crown and two stars below, the lieutenant with a solitary star, to the second lieutenant without any. The same basic styles exist today, with only minor variations. Nazi armed forces used their shoulder straps to carry marks of rank as well as devices on the collars of their tunics and coats. Specialist badges of the Nazi forces were worn on the left sleeve, and full lists of these will be found in the appropriate reference books.

Buttons form another field of collecting in which items are both cheap and readily available. Originally buttons were merely decorative and were of ivory, gold, silver and other precious metals, but by the late Tudor times button making was a well-established industry, and by 1660 Birmingham was becoming one of the chief centres of production.

By 1680 brass buttons were being produced by the mould process, whereby thin sheets of brass were shaped and hammered to fit over a wood or bone former. Later it was forbidden to use ordinary metal buttons apart from those of gold and silver, but this restriction was lifted in 1741, and from then on metal buttons held the field. Gilt buttons made from a brass alloy were produced from the 1750s, and a good gilding process was introduced around 1790. For cheaper buttons, pewter, tin and brass were used, and there are frequent references to such materials being used for military buttons during the late 17th and early 18th centuries. Clothing records of various military units suggest that button design was a matter for the whim of the Colonel, but in September 1767 it was decreed by the British War Office that military buttons were to bear the number of the Regiment – a practice in France from some five years earlier. The three regiments of Dragoon Guards were a little later accorded the right to have their initials in place of a regimental number.

At first most of the buttons were flat, but became more and more convex until 1855, when a rimmed button was introduced. Pewter buttons for other ranks were replaced in the same year by brass ones. In 1871 it was decreed that buttons bearing the regimental design were no longer to be worn by other ranks. In their place most regiments were issued with a button bearing only the Royal Arms, but in 1928 the regimental pattern was reintroduced.

Some guidance in dating buttons is offered by the stamping of the manufacturer's name on the back – a practice which first started around the 1800s to 1820. Since only a handful of manufacturers supplied the majority of buttons and these firms were in production for many years, it has been possible to date the changes in name, so that the style of name gives a broad guide to the date of manufacture. Parkyn's book on *Shoulder Belt Plates and Buttons* gives a full list. Further indication in the case of very old buttons is given by the method of attachment of the shank to the back of the button. Around about 1770 a single wire loop brazed at one point was used, and by 1780 the loop was being attached by two flat extensions. The shank was made of hand-drawn wire which may be recognised by its variations in thickness. By 1840 the shank was being produced automatically.

Identification is usually very straightforward, and once again Parkyn's book and Edwards' *Regimental Badges* will cover most buttons. It is usually apparent which buttons are military, but it is perhaps possible to mistake a livery button, normally worn by servants, and *The Complete Button Book*, by L. Albert and K. Kent, will prove useful.

Buttons of the last half century are still plentiful, and it is possible to purchase single ones for a few coppers. Early examples are naturally very scarce and fetch varying amounts. However, they are by no means unobtainable. As with medals and badges, perseverance is essential, and every box of odds and ends should be examined closely.

Cleaning buttons and badges is very much a matter of common sense, and the same remarks are applicable to these as to medals. Rubbing should be kept to a minimum, and it may be preferable to lacquer them. Display is very much a matter of personal preference, but one simple style is to mount them on an appropriate background of card or hardboard by making small holes to hold the shank and securing them by a metal pin, nail or even a matchstick. For beginners, badges and buttons offer a very good introduction to the study of military history, and their availability has much to recommend them.

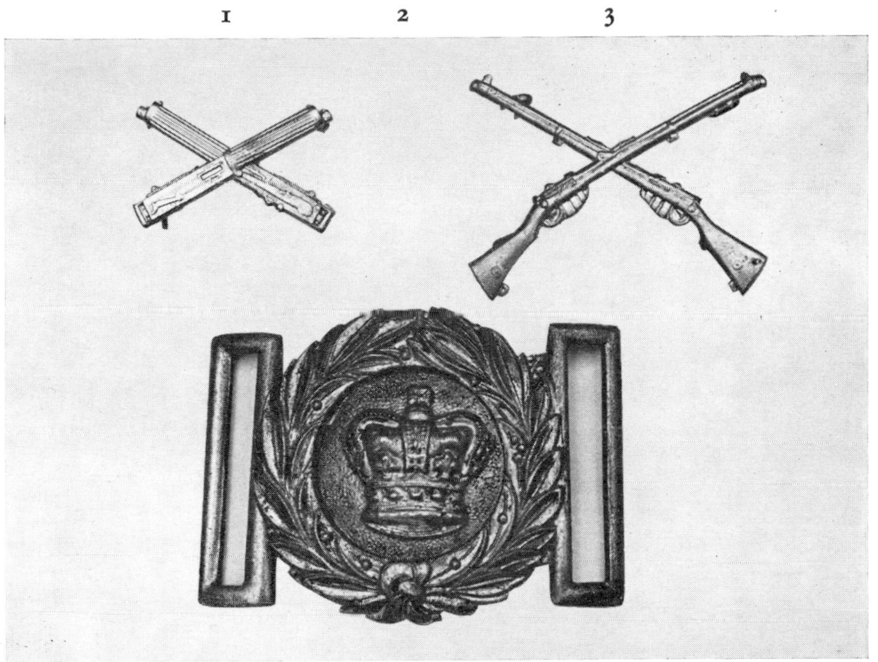

(1) Machine gunner's badge. (2) British military belt buckle. (3) Crossed rifles worn by a marksman.

Part of a brass belt buckle of the Prince of Wales Volunteers; late 19th century.

Nazi wound badge in silver, 1939. This type of badge dates back to the First World War, and was given to anyone who had received several wounds or a single serious wound.

Solid brass belt buckle of the Soviet Navy, hence the anchor bearing the star and hammer and sickle.

1 2 3

THE DUKE OF WELLINGTON'S (WEST RIDING REGIMENT)

Helmet Plate, 1870

Officer's Cap Badge, 1830

(1) Helmet plate of 1870. (2) Badge of the Duke of Wellington's (West Riding) Regiment. (3) Officer's cap badge of 1830.

▲
French gorget, dating from the Second Empire 1852–70, in brass wi
a silver eagle attached by three small nuts.

◀
German helmet plate of the Grand Elector's Royal (Leib) Cuirassiers (Silesian) No. 1.

1 2 3

(1) French Harbour Fire Defence badge.
(2) Canadian Cap General Service badge.
(3) Dutch Field anti-aircraft badge.

Selection of Indian Army badges of the Second World War. Most are self-explanatory; centre right is of the Indian Engineers, and the bottom centre is of the 27th Lancers, which was disbanded in 1948.

1 2

CANADIAN
SCOTTISH
SEMPER.
THE
CANADIAN
SCOTTISH
DEAS
CATH
CALGARY HIGHLANDERS
CAPE·BRETON·HIGHLANDERS
SIOL NA FEAR FEARAIL

3 4 5

◀

Many Scots emigrated to Canada during the 19th century, and the Canadian Army was rich in groups emphasising their Scottish origins: (1) Small badge of Canadian Scottish. (2) Badge of Wentworth Regiment. (3) Cap badge of Canadian Scottish. (4) Cap badge of Cape Breton Highlanders. (5) Cap badge of Calgary Highlanders.

▼

(1) Cap badge of the Fife and Forfar Yeomanry. (2) Collar badge of the West Kent Yeomanry. (3) Cap badge of Canadian Newfoundland Regiment.

1 2

3

I 2

3 4 5

(1) and (2) Australian Commonwealth Military Forces general service badges, the 2nd being cast in a dull bronze finish. (3) Symbol for physical training instruction of the British Army. (4) Imperial Service Badge. (5) Cap badge of Machine Gun Corps.

Cap badge of the 15th Canadian Battalion, 48th Highlanders.

Grenadier Guards officer's shoulder belt plate, *c.* 1837–55.

Ornate helmet plate of the Calcutta Volunteer Rifles.

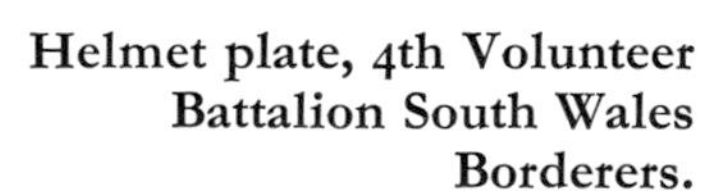

Helmet plate, 4th Volunteer Battalion South Wales Borderers.

I

(1) Cap badge of the Royal Northumberland Fusiliers. (2) Epaulette letters for Royal Horse Artillery 1914–18. (3) Brass lettering for epaulettes of the Canadian group, the 50th Gordons.

Infantry Officer's shako plate worn *c.* 1811–16.

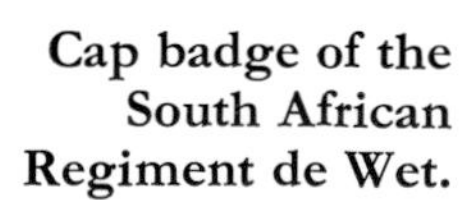

Cap badge of the South African Regiment de Wet.

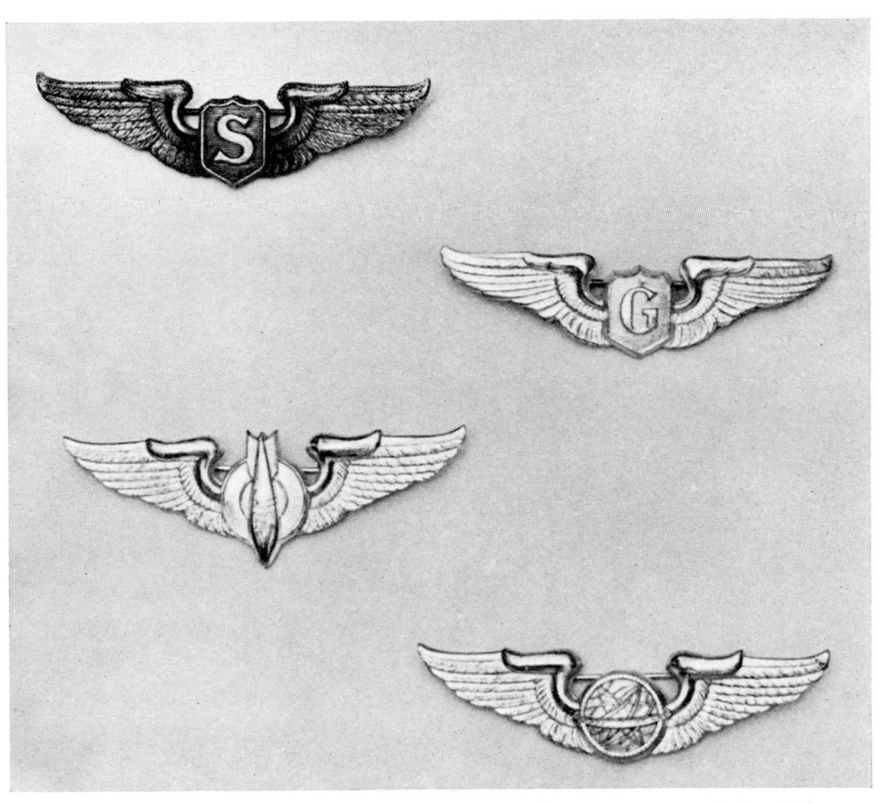

Selection of U.S. Army Air Corps breast badges, Second World War.

US
PLURIBUS
UNUM

▲ *Left:* General Service Corps cap badge. *Right:* A modern casting of part of the royal arms, similar to a Victorian helmet plate.

▶ Cap badge of a Soviet Russian Naval Officer. The star is in red, gold and white enamel.

◀ Selection of U.S. Second World War cap badges and buttons.

Left: Royal Marine Light Infantry helmet plate, 1879–1901.
Right: Royal Artillery helmet plate, 1879–1901.

British bosun's whistle and two naval buttons, both unmarked.

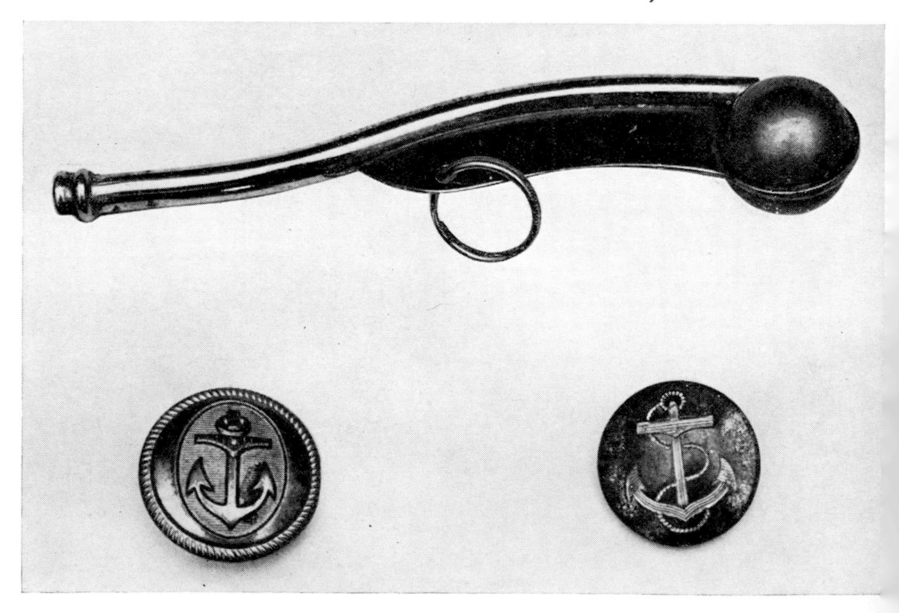

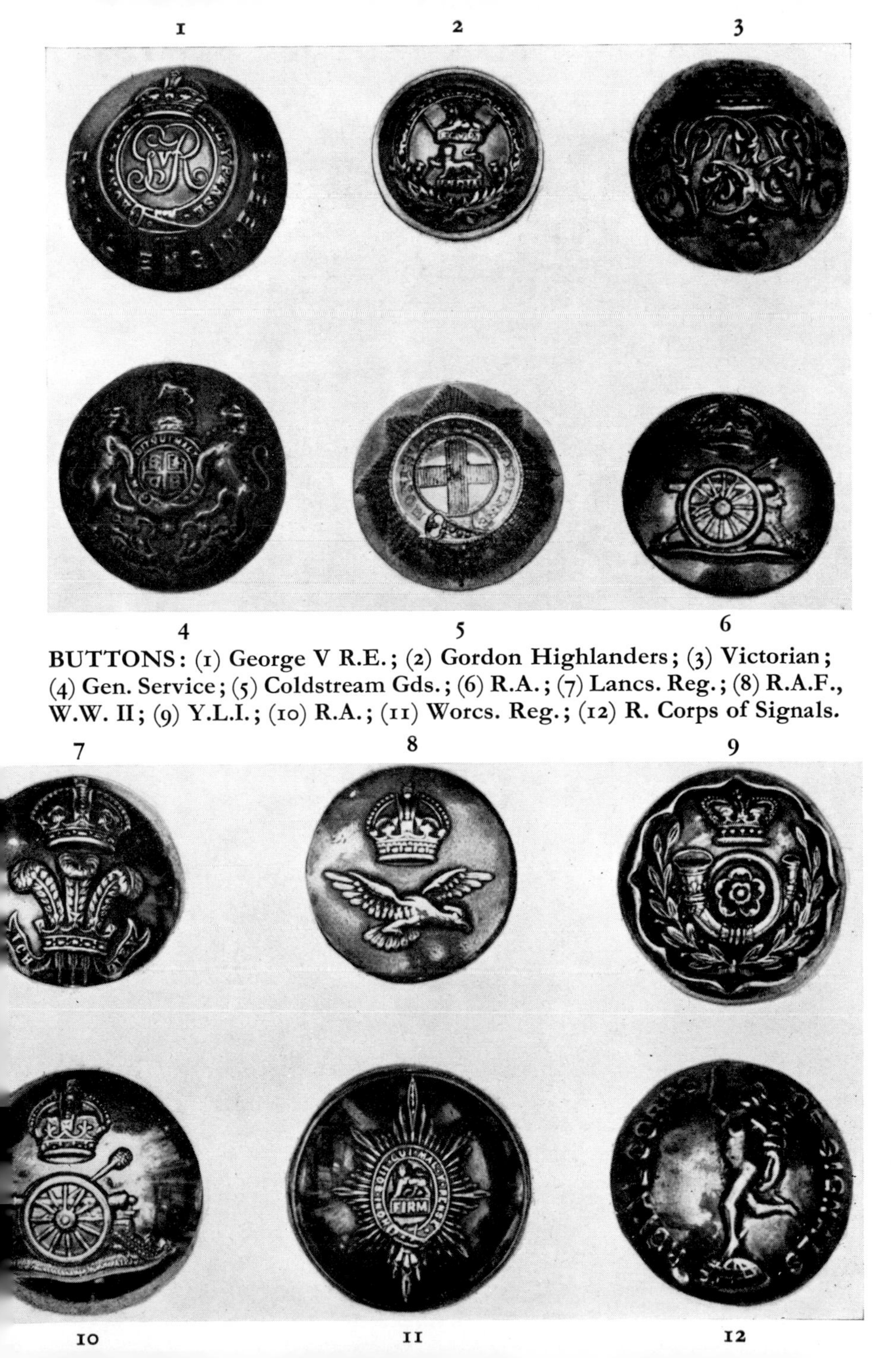

BUTTONS: (1) George V R.E.; (2) Gordon Highlanders; (3) Victorian; (4) Gen. Service; (5) Coldstream Gds.; (6) R.A.; (7) Lancs. Reg.; (8) R.A.F., W.W. II; (9) Y.L.I.; (10) R.A.; (11) Worcs. Reg.; (12) R. Corps of Signals.

1 2 3

4 5 6

BUTTONS: (1), (2), (3) and (5) Livery buttons bearing sundry coats of arms and emblems; (4) Staff Off.; (6) Merchant Navy; (7) R.N.; (8) Georg V; (9) R. Scots Reg.; (10) 5th R. Innisk. Drag. Gds.; (11) R.A.O.C.; (12) Norfolk Reg.

7 8 9

10 11 12

BUTTONS: (1) Welch Reg.; (2) Indian Army; (3) Queen's R. W. Surrey Reg.; (4) E. Surrey Reg.; (5) P.O.W. Leinster Reg.; (6) 5th R. Innisk. Drag. Gds.; (7) Princess of Wales' Own; (8) R.M.; (9) Duke of Cornwall's L.I.; (10) Royal Ordnance Corps; (11) Welsh Gds.; (12) 9th Battalion H.L.I.

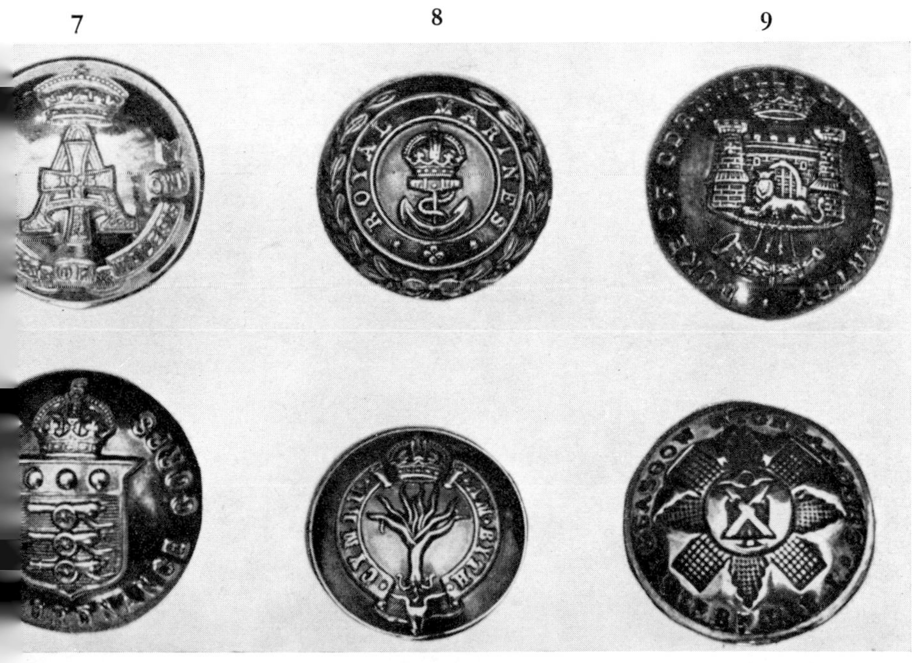

Police of the World, detail of colour plate from *Boys' Own Paper*, *c.* 1900.

IX

LAW AND ORDER

The story of the police in Britain is long and complex. Originally the onus of peace preservation had been placed on small local groups which were held directly responsible to the king or his officers for any crime committed within their district. It was not until the 16 and 17th centuries that the system incorporating the local constable was fully developed. This most unfortunate individual was elected annually by the members of the village or parish, and this elective system is essentially the same as that which still pertains today in much of the United States of America, where the local Sheriff is chosen by the inhabitants of a district. His powers, however, are considerably greater than any that the local constable ever possessed. The task of the constable was indeed a thankless one, and every effort was made to avoid this onerous task. For the unlucky individual who was finally chosen there was a need to invest in him some brief authority which could be recognised by the parishioners or villagers. This most often took the form of a tipstaff or truncheon. These vary considerably in size and shape, and often bear the name of the hundred, that is the local division of the district, the parish or the church. It is not always easy to identify examples, for in many cases the title was abbreviated and initials only were used. The Royal Arms were frequently included as part of the decoration, and this does give some limited guidance in dating specimens. Over the centuries the British Royal Arms have varied somewhat in detail, and these variations can be dated with accuracy, thus a particular styled coat of arms can provide two dates defining the period within which the example must have been made.

Some of the tipstaves were of metal, brass being the most popular, and in a few cases some were of silver. The tipstaff most commonly used in England is of brass and ebony, with a threaded crown at one end. It is very often said that the constable carried his warrant inside the tipstaff, but this, in fact, was not so. The great majority of tipstaves are solid, and even those which are hollow have a space too small to hold a warrant.

In London there was established in 1792 a small paid police force. Eight offices were set up with eight to twelve paid constables at each under the control of two or three magistrates. These eight offices were allocated government money to purchase their supplies and equipment, and it was therefore at the discretion of the clerk or magistrates as to the type of tipstaff or truncheon adopted. Truncheons or tipstaves from these police offices do occasionally turn up, and they will always bear the name of the office of origin, sometimes the initials, sometimes the full name, which may be Great Marlborough St.; Queens Square, Westminster; Hatton Garden; Worship St.; Whitechapel; Shadwell; Union Hall, Southwark; or Bow Street.

If it can be said that there exists any discernible line of development in the design of tipstaves it is one that shows a decrease in ornamentation as the 18th and 19th centuries proceed, until their use was abandoned around the middle of the 19th century. For a brief period a tipstaff was carried by the inspectors of the Metropolitan Police Force, but the practice was abandoned in the 1870s. Other officers apart from those of the police carried tipstaves as badges of office, and examples may be found bearing such legends as "Surveyor of the Highways", "Bank of England", "Overseer of the Poor" and sundry similar posts.

Truncheons or batons were used by most police forces in England, and are still carried by a number of American and Continental forces. Many of them combined the function of weapon and tipstaff, and frequently bore the Royal Arms as well as the title or name of the district and office. Some will be found marked S.C. or Special Constable, and these were issued to volunteers who joined the police to augment it during times of national or local crises. The shape of truncheons varies enormously, ranging from straight narrow sticks to thick barrel-shaped bludgeons. Some are actually pivoted at the centre to form a vicious flail.

When the Metropolitan Police was formed in 1829 a great deal of time and trouble was spent in selecting their equipment. Commissioners Mayne and Rowan and the Home Office were anxious that no charges of militarism could be directed against them, and therefore the uniform was deliberately chosen to look as unmilitary as possible. It comprised a top hat, specially strengthened at the crown, blue frock-coat and light-coloured trousers. Normally the only weapon carried was the truncheon, and this fitted into a tubular leather case suspended at the back of a thick, leather waist belt. At first these truncheons were highly decorative following on the style of those used by the local constable, but in the interests of economy this practice was abandoned around the middle of the century and plain staves bearing the letters M.P., frequently burnt in, were substituted. It was about the same period that the commissioners called together representatives from most of the divisions of the Metropolitan Police Force to consider possible changes in their truncheons and the means of carrying them. After much discussion it was decided that truncheon cases should be abolished, and in their place trousers were specially adapted to take a single deep pocket on the right-hand side. Into this dropped the truncheon, leaving only the leather wrist thong showing at the top, and this is the system still in use today.

It was soon apparent, even to the most hardened opponents, that the Metropolitan Police Force was a most successful innovation, and it was planned to extend the system to the rest of the country. Parliament therefore passed legislation, at first permissive but later obligatory, enabling authorities to set up their own local forces. These forces tended, rather naturally, to follow the lead set by the London Police, and many of them adopted the plain truncheon in place of the previously elaborate tipstaff. Some, however, did retain a measure of individuality, having the name of the town painted on the truncheon. In certain cases the practice of painting the truncheons continued well into the 20th century, and examples bearing the arms and name of Birmingham have been recorded bearing dates as late as 1923.

It is unfortunate that the very nature of truncheons has meant that so many of them, indeed most of them, have, over the years, been rubbed, chipped, scratched or dented. Ideally only good-quality pieces are worth collecting, but there must be certain qualifications to this sweeping generalisation, and rarity must also be considered. In the case of the London Police Offices, where it is known that only a very limited number were issued, then one must be prepared to accept items of a lower standard than one might hope for. In the last few years replicas of these painted truncheons have been produced and, newness aside, it is not always easy to be certain as to which is the original and which the reproduction. Most of the reproduction truncheons have the Royal Arms in the form of a transfer. Unfortunately this, in itself, is not conclusive, for some of the earlier examples also made use of this system.

The display of truncheons is not an easy matter, although they are most attractive and decorative. Some collectors mount them in racks, and if this is done it is most important that the system used involves no danger of scratching the truncheons. One of the best, admittedly rather elaborate, means of display is the semicircular wooden arc mounted above a base. Holes are drilled through the top arc and the truncheon is dropped through this to rest on the base, so forming a fan-shaped display. The practice of driving screw eyes or hooks into one end of the truncheon to suspend it is not to be encouraged. Fortunately the majority of truncheons and tipstaves need little more attention than a good, careful cleaning. Since most of the painting on them is in oils, they can be washed with little risk of damage. Once cleaned, a gentle polish with one of the modern silicone products will ensure that the item stays in good condition with a minimum of attention. If they have to be stored together it is well to wrap them in newspaper or put them into a sock or similar container to prevent them rubbing or scratching.

Most of the truncheons were supplied by a handful of makers. During the first half of the 19th century William Parker and Parker Field, who continued the business until the 1870s, supplied a very large percentage, especially to forces in the south-east of England. The name and place of manufacture will usually be found stamped on the base of the truncheon. Actually this statement needs some clarification, for quite often the name found there is not that of the manufacturer but rather the retailer. During times of emergency, such as the Chartist riots during the first half of the 19th century, large numbers of truncheons were produced at very low costs, and often these have no identifying marks.

Prior to the formation of the Metropolitan Police Force in 1829, few police officers wore any kind of uniform apart from the Thames River Police and an earlier Horse Patrol. The top hat was replaced during the middle of the 19th century by the now familiar police helmet, and this, together with the cloth cap and the City of London pattern with its raised central cone, are still the most usual forms of headdresses in use today. In America the great majority of police officers wear either cloth caps or stetson-type hats. Europe offers a variety of patterns, but the sum total is probably not more than twenty. With this limited field, the majority of collectors concentrate on helmet plates and badges. Small groups of these turn up in the sale rooms, and odd examples may be found in the antique markets. For the keen collector it is often possible to obtain examples directly from the force concerned, as these authorities are usually quite co-operative.

Another very limited and specialised field is that of handcuffs or manacles. The majority of these offered for sale are army surplus and are quite modern, and most bear the manufacturer's name and date of supply. The earlier forms of handcuffs and leg irons were riveted into position, and it is not until the late 18th and early 19th centuries that locks were instituted. Early manacles were sometimes fitted with a hasp and loop to take a small padlock, and later came the screw type with a key which was inserted in one end and engaged with a threaded pillar. As the key was rotated the pillar was withdrawn against the pressure of a spring allowing the handcuff to be opened. Similar is the plug type of handcuff, and in these the opening to accomodate the key is covered by a screwed plug which must be removed before the handcuff can be unlocked. These earlier types are less common, but by no means rare. In addition to the locking handcuff there is the spring-loaded snip used to secure the offender for a short journey. These are not fitted with locks, but are of the figure-of-eight pattern, one loop encircles the wrist of the offender the other is gripped firmly by the escorting officer.

These, then, are the main fields for collecting police items, but there are many others which may offer less scope, such as general equipment, lamps, whistles, uniforms, buttons or, far rarer, posters and documents. The collector of police material will always encounter a certain amount of difficulty in positive identification, for the amount of printed material dealing with the detailed history of equipment is very limited. However, there is a growing interest in this field, and it is to be hoped that this state of affairs will be remedied in the not too distant future.

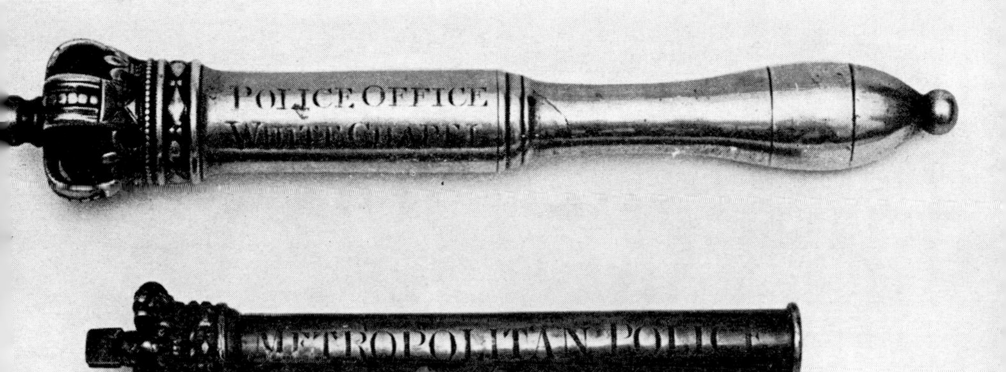

Top: Engraved brass tipstaff intended as a means of identification for police of the Whitechapel Police Office, set up in 1792. *Bottom:* Engraved brass tipstaff; identification for plain clothes Metropolitan policemen issued after the force was organised in 1829.

Rattle of the type carried by the Watch and by early police forces.

The Metropolitan Police Force, after its initial success, often helped to organise local forces. *Left:* Helmet badge of the Oldham Police Force. *Right:* Helmet plate of Barnsley Borough Police.

Arm band with aluminium plate worn by civilian policemen enrolled in special emergencies, a practice begun in the early 19th century.

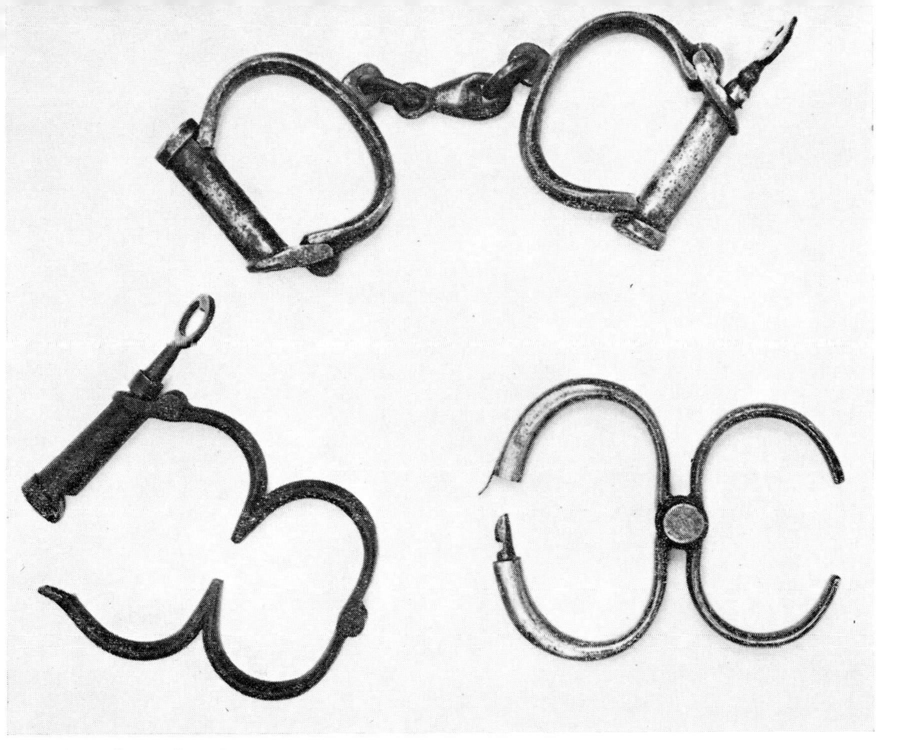

Handcuffs and snips
dating from the 19th century.

Example of a 19th-century
figure-of-8 handcuff,
unlocked by means of a
screw key and lacking any
central length of chain.
The leather case attached
to a belt.

MILITARY

ANTIQUITIES

RESPECTING

A HISTORY

OF

THE ENGLISH ARMY

FROM

THE CONQUEST TO THE PRESENT TIME.

VOL. I.

By Francis Grose Esq.r F.A.S.

LONDON

PRINTED FOR S. HOOPER No. 212 HIGH HOLBORN.

MDCCLXXXVI.

Title page of Francis Grose's *Military Antiquities;* 1st ed. of 1786.

X

BOOKS AND COLLECTIONS

The unfortunate prevalence of war is indicated by the vast number of books devoted to its every aspect, and the collector will soon find that it is extremely difficult to acquire all these books dealing with any particular topic. Many of the most useful books have long been out of print, and copies have acquired an antiquarian value in their own right. A few have been reprinted, and it may well be that others will be reprinted in the future, but still more will remain scarce and difficult to obtain. A glance at A. S. White's *Bibliography of Regimental Histories* will soon make clear the number of books devoted to the British Army alone. Many regimental histories are unbelievably dull and frequently contain little really useful information on uniforms, weapons and organisation, but deal only with the campaigns and battles or social activities. While much of this information may be of use to the military historian, it has little value for the collector. Nevertheless there are often odd items of information buried in the lifeless prose, and such books should always be checked. Others, such as Macdonald's *History of the Dress of the Royal Artillery*, abound in useful and generally reliable information on uniforms and equipment.

The recent increase in interest has produced a large number of new and scholarly works, and details of these are usually to be found in the magazines and journals listed below.

For the collector who wishes to trace any particular member of the services, the Army, Navy and Air Force Lists and U.S. Congressional Records are invaluable. In a few cases fairly full information on officers and men is available as far back as the 17th century. For certain periods, works such as Clowes' *Naval Biography* will supply pretty full details of the service of officers. In the case of officers it is always worth checking the *Dictionary of National Biography*, for the subject may rate an entry which will indicate follow-up lines of enquiry.

If all other sources prove useless, then enquiries may be sent to the appropriate War Office or Regimental Archives, but if this proves necessary, then all the known details must be given.

Although the great majority of the books listed below are in English, it must be remarked that most national bibliographies include a number of book titles devoted to the study of their own armed forces. France, in particular, is well served. A number of national museums also issue journals, magazines and catalogues devoted to the study of military history in the broadest sense.

The books listed below have been classified, as far as possible, to correspond broadly with the sections of this book, but in many cases one book may well cover a number of topics. This bibliography is limited in scope, and there will be many titles that other collectors may feel should have been included, but the field is so vast that some selection has to be made.

Books out of print or otherwise difficult to obtain are marked with an asterisk, but a mention is useful, since on many occasions loose prints from these books may be found and it is important to be able to identify the source and date.

GENERAL COLLECTING

Chats on Military Curios, A. C. Johnson. London, 1915

PRINTS

**British Military Prints*, R. Neville. London, 1909
**Chats on Old Prints*, A. Hayden. London, 1906

Military Drawings and Paintings in the Collection of H.M. The Queen, A. Haswell and N. Dawnay. Vol. 1. London, 1966
**Print Restoration and Picture Cleaning*, M. J. Gunn. London, 1911
**Prints of British Military Operations 1066–1868*, C. de W. Crookshank. London, 1921

MEDALS

British Battles and Medals, L. L. Gordon. Aldershot, 1962
Coronation and Commemorative Medals, H. H. Cole. Aldershot, 1953
**Men Whose Fathers Were Men*, "Centurion". London, 1925
Orders, Medals and Decorations, P. Hieronymussen. London, 1967
Ribbons and Medals, H. T. Dorling. London, 1946
**War Medals and Decorations 1588–1898*, D. Hastings Irwin. London, 1899

UNIFORMS

Arms and Equipment of the Civil War, Jack Coggins. New York, 1962
Army Uniforms of the World, F. Blakeslee. Hartford, U.S.A., 1919
Anatomy of Glory, W. Lachouque and A. S. Brown. London, 1962
British Military Uniforms from Contemporary Pictures, W. Y. Carman. London, 1957; Reprint 1968
**Costumes de Guerre*, Paris, 1882
Cavalry Uniforms of British Army, P. Smitherman. London, 1962
Dix Siecles de Costume Militaire, H. Lachouque. Paris, 1963
Five Centuries of American Costume, R. Turner-Wilcox. London, 1963
Handbuch der Uniformkunde, H. Knotel and H. Sieg. Hamburg, 1937
**History of the Dress of the British Soldier*, J. Luard. London, 1852
**History of the Dress of the Royal Regiment of Artillery 1625–1897*, R. J. Macdonald. London, 1899

History of Regiments and Uniforms of the British Army, R. M. Barnes. London, 1957

History of Uniforms of The British Army (5 vols.), C. P. Lawson. London, 1940–67

Indian Army Uniforms, W. Y. Carman. London, 1961; Reprint 1968

Infantry Uniforms of the British Army, P. Smitherman. London, 1965

Le Costume et les Armes des Soldats de Tous les Temps, L. and F. Funcken. Paris, 1966

Military Costume, P. Martin. London, 1967

Military Uniforms of Britain and the Empire, R. M. Barnes. London, 1960

Ranks and Uniforms of the German Army, Navy and Air Force, D. Irlam. London, 1939

Red Army Uniforms and Insignia 1944. London, Reprint 1968

**Uniformes*, F. J. Dally. Paris, 1886

**Uniformes de l'Armée Allemande*, F. J. Dally. Paris, 1885

Uniforms and Dress of the Army and Navy of the Confederate States of America, Ed. R. Harwell. Reprint, Philadelphia, 1960

Uniforms and Insignia of the Third Reich, S. Hyatt. U.S.A., 1962

Uniforms of the Royal Artillery, P. Smitherman. London, 1966

Uniforms of the Scottish Regiments, P. Smitherman. London, 1962

**Uniforms of the World*, F. Blakeslee. New York, 1929

Uniforms of the Yeomanry Regiments, P. Smitherman. London, 1967

BOOKS

Bibliography of English Military Books up to 1642, M. J. Cockle. London, Reprint 1960

Bibliography of Regimental Histories of the British Army, A. S. White, London, 1965

**Catalogue of Books Relating to Military History of India*, M. J. Cockle. Simla, 1901

MODEL SOLDIERS AND WAR GAMES

Charge, P. Young and J. P. Lawford. London, 1967
Collecting Toy Soldiers, J. Nicollier, Rutland, U.S.A., 1967
**Floor Games*, H. G. Wells. London, 1911
**Little Wars*, H. G. Wells. London, 1913
Model Soldiers, J. G. Garratt. London, 1962
Model Soldiers, E. Harris. London, 1962
Tackle Model Soldiers this Way, D. F. Featherstone. London, 1963
War Games, D. F. Featherstone. London, 1962

WEAPONS

American Socket Bayonets, D. B. Webster. Ottawa, 1964
Bayonets, F. J. Stephens. London, 1968
British Artillery, S. J. Gooding. Ottawa, 1965
British Military Bayonets, R. W. Latham. London, 1967
British Military Firearms, H. L. Blackmore. London, 1961
British Military Swords, J. W. Latham. London, 1966
**Engines of War*, H. Wilkinson. London, 1841
**Exercise for Broadsword*, B. Wayne. Washington, 1850
**Exercise of Firelock*. London, 1712
French Military Weapons, J. E. Hicks. U.S.A., 1964
German Infantry Weapons, Ed. D. B. McLean. U.S.A., 1967
**Hungarian and Highland Broad Sword*, T. Angelo. London, 1799
Manual of Platoon Exercises, T. L. Mitchell. Ottawa, 1965
Monographies de l'Arme Blanche (1789–1870) et de l'Arme à Feu Portative (1718–1900), M. Bottet. Paris, 1959
**Our Engines of War*, H. J. Jervis. London, 1859
Small Arms, F. Wilkinson. London, 1965
Small Swords and Military Swords, A. V. Norman. London, 1967
Swords and Daggers, F. Wilkinson. London, 1967
Sword, Lance and Bayonet, C. J. Foulkes. London, Reprint 1967
The British Soldier's Firearms 1850–64, C. H. Roads. London, 1964
Weapons and Tactics, J. Weller. London, 1966

Weapons of the American Revolution, C. C. Neumann. New York, 1967
Weapons of the British Soldier, H. C. B. Rogers. London, 1960
Weapons of War, T. Cleator. London, 1967

BADGES, BUTTONS, ETC.

Badges and Emblems of the British Forces 1940. London, Reprint 1968
Badges on Battledress, H. N. Cole. Aldershot, 1953
Heraldry in War, H. N. Cole. Aldershot, 1953
**Old Scottish Regimental Colours*, A. Ross. Edinburgh, 1885
**Rank and Badges*, O. L. Perry, London, 1887
**Records and Badges of the British Army*, H. M. Chichester and G. Burgess Short. London, 1900
Regimental Badges, T. Edwards. Aldershot, 1966
**Regimental Records*, J. S. Farmer. London, 1901
Shoulder Belt Plates and Buttons, M. Parkyn. London, 1956

POLICE

**Truncheons*, Erland Fenn Clark. London, 1935
**The History of Truncheons*, E. R. H. Dicken. N. Devon, 1952
**The Rise of Scotland Yard*, Douglas G. Browne. London, 1956
The World's Police, James Cramer, London, 1964
**Truncheons and Tipstaves*, Charles Stuart. (Vol. 5 of the *Concise Encyclopaedia of Antiques*.) London, 1961

CAVALRY

**A Travers l'Europe Croquis de Cavalerie*, L. Vallet. Paris, 1893
British Regular Cavalry, L. Cooper. London, 1965
**Cavalry, L. Nolan*. London, 1853
**Guns and Cavalry*, E. May. London 1896,
**History of Chivalry*, F. Lottenkamp. London, 1857
Horses and Saddlery, G. Tylden. London, 1965
**La Cavalerie Français*, H. Choppin. Paris, 1893

Le Chic à Cheval, L. Vallet. Paris, 1891
Mounted Troops of the British Army 1066–1948, H. C. Rogers. London, 1959

NAVAL

Biographia Navalis, J. Charnock. London, 1794–97
British Navy Book, C. Field. London, 1915
Dress of Naval Officers, W. E. May. London, H.M.S.O., 1966
Naval and Military Costumes, J. Atkinson. London, 1807
Naval Biographical Dictionary, W. O'Byrne. London, 1849
Naval Officer's Sword, H. T. Bosanquet. London, H.M.S.O., 1958
Naval Swords and Firearms, W. May and A. Kennard. London, H.M.S.O., 1962
Royal Naval Biography, J. Marshall. London, 1823–33
Royal Navy, W. L. Clowes. London, 1897–1903
Uniforms of the Navy Past and Present, W. T. Carman. London, 1942
Uniforms of the Royal Navy, J. Mollo. London, 1965

GENERAL HISTORIES

Armed Forces of South Africa, G. Tylden. Johannesburg, 1954
Armies of Asia and Europe, E. Upton. New York, 1878
Armies of Europe, Count Gleichen. London, 1890
Armies of India, C. Lovell and C. MacMunn. London, 1911
Armies of the World, C. S. Jerram. Paris, 1899
Art of Warfare in Biblical Lands, Y. Yadin. London, 1963
Army Book, J. C. Dalton. London, 1893
Battlefields of Europe, Vol. 1, *Western Europe*, D. Chandler. London, 1965
Battlefields of Europe, Vol. 2, *Central and Eastern Europe*, D. Chandler. London, 1965
Battles of the 19th Century, A. Forbes and Others. London, 1896
British Army Book, C. Field. London, 1915

Confederate Military Land Units 1861–65, Compiled by W. J. Tancig. London and New York, 1967
**Cromwell's Army*, C. H. Firth. London, 1902
**Famous Regiments of British Army*, W. H. Davenport Adams. London, 1864
**George I's Army*, C. Dalton. London, 1910
**Historical Records of the British Army* (70 Histories), Ed. R. Cannon. London, 1835
**History of the Art of War in the Middle Ages*, C. Oman. London, 1924
**History of the British Army* (13 vols.), Sir J. W. Fortescue. London, 1899–1930
**History of the British Standing Army*, C. Walton. London, 1894
History of Scottish Regiments, W. P. Paul. London, 1967
History of the Soviet Army, M. Garder. London, 1966
**History of the U.S. Army*, W. A. Ganoe. New York, 1924
**Indian Army*, E. H. Collen. Oxford, 1907
**La France Militaire*, F. J. Dally. Paris, 1885
**L'Armée Danoise*, F. J. Dally. Paris, 1888
**L'Armée Française*, H. L. Choppin. Paris, 1890
**Les Armées Etrangères*, F. J. Dally. Paris, 1885
**Military Antiquities*, F. Grose. London, 1796
Regimental History of Cromwell's Army, Sir C. Firth. Oxford, 1940
**Scottish Regiments of the British Army*, Ed. I. H. Mackay Scobie. London, 1942
Soldiers and Armies, E. E. Tucker. London, 1967
**The Book of the Thin Red Line*, Sir H. Newbolt. London, 1915
**The British Army*, Sir S. D. Scott. London, 1868
**The British Army*, Lt. Cmd. E. C. Talbot-Booth.
The British Army 1642–1970, P. Young. London, 1967
The British Soldier, H. de Watteville. London, 1954
War, E. Sheppard. London, 1967
Warrior to Soldier, A. V. Norman. London, 1966

PERIODICALS DEVOTED ENTIRELY TO MILITARY MATTERS

There were a number of magazines published during the 18th and 19th centuries devoted entirely to military and naval matters. Many of these lasted for very short periods, others continued for longer periods and some were amalgamated. Many of these publications turn up in booksellers' lists and in bookshops. Some are very useful, others are full of rather dull and tedious articles. It is always worth checking these, and some of the more common ones are listed below.

Army and Navy Magazine
The Cavalry Journal
Journal of the Royal United Services Institution
Illustrated Naval and Military Magazine
Military Annual (1844)
Military Magazine (1793)
Military Panorama (1812)
Military and Naval Medal Magazine (1896)
Naval and Military Magazine (1827)
Proceedings of the Royal Artillery Institution
The Royal Military Chronicle

PERIODICALS DEALING WITH GENERAL MILITARY MATTERS

Journal of the Society for Army Historical Research
Soldier (H.M.S.O., London)
The Bulletin of the Military Historical Society (London)
Tradition (London)

PERIODICALS CONTAINING FREQUENT ARTICLES ON MILITARY MATTERS

Waffen-und Kostumkunde (Munich and Berlin)

Armi Antiche (Turin)
Vaabenhistorishe Aarbøger (Copenhagen)

WAR GAMES

There are a number of newsletters and pamphlets issued by Societies. Although primarily concerned with War Games, they often contain articles dealing with general military topics.

Airfix Magazine (London)
Miniature Parade (California, U.S.A.)
Miniature Warfare (London)
Sling Shot (Southampton)
Strategy and Tactics (*Journal of American War Gaming*) (New York)
The Despatch (Argyle)
War Games News Letter (Southampton)

MILITARY MINIATURES

Societies devoted to the interests of collectors are to be found in many countries, and such Societies frequently issue periodicals. Again, although aimed mainly at those with a particular interest in miniatures, many of the articles are concerned with general military matters.

A very full list of these Societies is included in *Model Soldiers*, by J. G. Garratt, and a list of the better-known suppliers of model figures is to be found in *Tackle Model Soldiers this Way*, by D. F. Featherstone.

COLLECTIONS OF MILITARIA

There are comparatively few museums devoted entirely to militaria. There are many small Regimental Museums housing collections of essentially local military interest, and many provincial museums have cases which contain items of local interest with a military connection. The list of Regimental Museums is very long indeed (See pages 233–249). Many items of military interest are also housed in major collections under special sections, such as metalwork, textiles, costume and similar broad headings. The collections listed below contain sufficient items of military interest as to make them worthy of a special visit.

GREAT BRITAIN

London	H.M. Tower of London
	Victoria and Albert Museum
	National Maritime Museum
	Rotunda, Woolwich
Sandhurst	National Army Museum
Enfield	Pattern Room (special permission needed)
York	Castle Museum
Glasgow	Art Gallery and Museum
Edinburgh	Scottish United Services Museum
	National Museum of Antiquities
	Royal Scottish Museum

AUSTRIA

Graz	Steiermarkisches Landeszeughaus
Vienna	Kunsthistorisches Museum
	Heeresgeschichliches Museum
	Historisches Museum der Stadt Wien

BELGIUM

Brussels	Musée de la Porte de Hals
	Musée Royal de L'Armée

CANADA

Kingston	Old Fort Henry
Toronto	Royal Ordnance Museum

DENMARK

Copenhagen	Tøjhusmuseet

FRANCE

Paris	Musée de L'Armée

GERMANY EAST

Berlin	Museum fur Deutsche Geschichte
Dresden	Historiche Museum

HOLLAND

Leiden	Het Nederlands Leger-en Wapenmuseum "General Hoefer"

ITALY

Florence	Museo Nazionale
Rome	Museo Nazionale di Castel S. Angelo
Turin	Armeria Reale
	Museo Nazionale Storico d'Artiglieria
Venice	La Sala d'Armi, Palazzo Ducale

NORWAY

Oslo	Haermuseet

POLAND

Cracow	Museum Narodowe Krakowie
Warsaw	Polish Army Museum

RUSSIA

Leningrad	State Hermitage
	Central Museum of the U.S.S.R. Navy

	Museum of the Engineer and Artillery Troops
Moscow	Kremlin, Armoury
	State Historical Museum

SPAIN

Madrid	Institutio de Valencia de Don Juan
	Museo del Ejercito Espanõl

SWEDEN

Stockholm	Kungl Livrustkammaren
	Kungliga Armemuseum

SWITZERLAND

Geneva	Musée d'Art et d'Histoire
Solothurn	Zeughaus
Zürich	Schweizerisches Landesmuseum

UNITED STATES

Georgia	Fort Oglethorpe	Chickamauga-Chattanooga National Military Park
	Fort Benning	U.S. Army Infantry Museum
Illinois	Rock Island	Rock Island Arsenal
Kansas	Larned	Fort Larned National Landmark
Louisiana	New Orleans	Louisiana State Museum
	New Orleans	Louisiana Historical Association
Maryland	Baltimore	Fort McHenry National Monument and Historical Shrine

	Aberdeen	U.S. Army Ordnance Center and School Museum
Massachusetts	Boston	First Corps Cadets Armory
	Salem	Essex Institute
	Springfield	Springfield Armory Museum
New York	New York	Metropolitan Museum
	Ticonderoga	Fort Ticonderoga
	West Point	West Point Museum
Oklahoma	Fort Sill	Fort Sill Artillery Museum
Texas	San Antonio	Witte Museum
	Fort Bliss	Fort Bliss Replica Museum
Virginia	Williamsburg	Colonial Williamsburg
	Richmond	Confederate Museum
	Quantico	United States Marine Corps Museum
Washington D.C.		Smithsonian Institution

POLICE MUSEUMS

Many forces have small collections of local interest. The largest collections of Tipstaves and Truncheons are probably those at The Castle Museum, York, and the Police College, Bramshill.

The Metropolitan Police Force are building up a useful museum at Bow Street, London.

APPENDIX

MILITARY MUSEUMS

LONDON

IMPERIAL WAR MUSEUM

Address: Lambeth Road, S.E.1.

Open: Monday to Saturday 1000 to 1800; Sunday 1400 to 1800.

Films from the Museum's collection are shown in the cinema Monday to Friday 1200; Saturday 1445, Sunday 1445 and 1600. Admission free.

MUSEUM OF ARTILLERY (THE ROTUNDA)

Address: Woolwich, S.E.18.

Open: Monday to Friday 1000 to 1245, 1400 to 1600; Saturday 1000 to 1200, 1400 to 1600; Sunday 1400 to 1600 (all days extended to 1700 April to September).

ROYAL ARTILLERY MUSEUM

Address: The Royal Military Academy, Woolwich S.E.18.

Open: Monday to Friday 1000 to 1230, 1400 to 1600.

ROYAL FUSILIERS MUSEUM

Address: H.M. Tower of London, E.C.3.

Open: Monday to Saturday 1000 to 1630; Sunday (summer only) 1400 to 1700.

BRIGADE OF GUARDS

The five regiments of Foot Guards do not have a museum at present, but hope to establish a Brigade of Guards museum in the near future.

THE MIDDLESEX REGIMENTAL MUSEUM

Address: R.H.Q. The Queen's Regiment (Middlesex Office), Deansbrook Road, Edgware, Middlesex.

Open: By appointment.

BERKSHIRE AND WESTMINSTER DRAGOONS MUSEUM

Address: Regimental Headquarters, 1 Elverton Street, Horseferry Road, S.W.1.

Open: By appointment.

21ST SPECIAL AIR SERVICE REGIMENT (ARTISTS) MUSEUM

Address: 17 Duke's Road, W.C.1.

Open: By arrangement only.

BEDFORDSHIRE

THE BEDFORDSHIRE AND HERTFORDSHIRE MUSEUM

Temporarily closed for move to a new location.

BERKSHIRE

HOUSEHOLD CAVALRY MUSEUM

Address: Combermere Barracks, Windsor

Open: Monday to Friday 1000 to 1300, 1400 to 1700; Sunday 1100 to 1300, 1400 to 1700.

THE ROYAL BERKSHIRE REGIMENT MUSEUM

Address: Brock Barracks, Oxford Road, Reading.

Open: Monday to Friday, but only on application to Regimental Headquarters.

THE R.E.M.E. MUSEUM

Address: Moat House, Arborfield, Reading.

Open: Monday to Friday 0830 to 1230, 1400 to 1700.

CHESHIRE

REGIMENTAL MUSEUM, THE CHESHIRE REGIMENT

Address: The Castle, Chester.

Open: Tuesday to Saturday and Sunday, 1030 to 1230, 1400 to 1600 (all extended to 1800 in summer).

REGIMENTAL MUSEUM, 3RD CARABINIERS (PRINCE OF WALES'S) DRAGOON GUARDS

Address: The Dale, Liverpool Road, Chester.

Open: Monday to Friday 0900 to 1700.

CORNWALL

DUKE OF CORNWALL'S LIGHT INFANTRY REGIMENTAL MUSEUM

Address: The Keep, The Barracks, Bodmin.

Open: Monday to Friday 0900 to 1230, 1400 to 1645. Saturday 1000 to 1230.

CUMBERLAND

THE BORDER REGIMENT MUSEUM

Address: The Castle, Carlisle.

Open: Monday to Saturday (March to April) 0930 to 1700; Sunday (March to April) 1400 to 1730; Monday to Saturday (May to September) 0930 to 1900; Sunday (May to September) 0930 to 1900; Monday to Saturday (October) 0930 to 1730; Sunday (October) 1400 to 1730; Monday to Saturday (November to February) 0930 to 1600; Sunday (November to February) 1400 to 1600.

DEVON

THE DEVONSHIRE REGIMENT MUSEUM

Address: Wyvern Barracks, Exeter.

Open: Monday to Friday 0900 to 1700 (at other times by special arrangement).

DORSET

ROYAL ARMOURED CORPS TANK MUSEUM

Address: Bovington Camp, near Wareham.

Open: Monday to Friday 1000 to 1230, 1400 to 1645; Saturday, Sunday and bank holidays 1030 to 1230, 1400 to 1600.

DORSET MILITARY MUSEUM

Address: The Keep, Dorchester.

Open: Monday to Friday 0900 to 1700; Saturday (October to June) 0900 to 1200; Saturday (July to September) 0900 to 1700. Admission 1*s*. (children 6*d*.). Conducted tours for parties by appointment.

MUSEUM OF THE ROYAL CORPS OF SIGNALS

Address: The School of Signals, Blandford Camp, Blandford Forum.

Open: Monday to Friday 1000 to 1230, 1400 to 1700; Saturday 1000 to 1200.

DURHAM

THE DURHAM LIGHT INFANTRY MUSEUM

Closed – collection, at present in store, has been loaned to Durham County Council and will eventually form part of a new folk museum at present being built.

ESSEX

THE ESSEX REGIMENT MUSEUM

Address: Eagle Way, Warley, Brentwood.

Open: Monday to Friday 1000 to 1200, 1400 to 1600. Saturdays and Sundays by appointment.

GLOUCESTERSHIRE

MUSEUM OF THE GLOUCESTERSHIRE REGIMENT (GLOUCESTER CORPORATION)

Address: 103 Westgate Street, Gloucester.

Open: Monday, Tuesday, Wednesday, Friday, Saturday 1000 to 1230, 1330 to 1715; Thursday 1000 to 1245.

11TH HUSSARS

Museum not yet open.

HAMPSHIRE

AIRBORNE FORCES MUSEUM

Address: Maida Barracks, Aldershot.

Open: Monday to Friday 0900 to 1630 (other days and times by application to Curator).

REGIMENTAL MUSEUM, ROYAL CORPS OF TRANSPORT

Address: Queen Elizabeth Barracks, Crookham.

Open: Monday to Friday 1000 to 1215, 1415 to 1630 (excluding public holidays). Advance warning preferred for parties of more than ten.

THE ROYAL GREEN JACKETS MUSEUM

Address: Peninsula Barracks, Winchester.

Open: Monday to Friday 1000 to 1600. Saturday (April to September) 1000 to 1200.

(Incorporates museums of the 43rd and 52nd, The King's Royal Rifle Corps and The Rifle Brigade).

R.A.M.C. HISTORICAL MUSEUM

Address: Keogh Barracks, Ash Vale, Aldershot.

Open: Monday to Friday 0900 to 1700; weekends by appointment.

THE ROYAL HAMPSHIRE REGIMENTAL MUSEUM

Address: Searle's House, Southgate Street, Winchester.

Open: Monday to Friday 1000 to 1230, 1400 to 1600 (except public holidays).

10TH ROYAL HUSSARS

Hoping to open a museum

ROYAL ARMY PAY CORPS MUSEUM

Address: Worthy Down, Winchester.

HEREFORDSHIRE

HEREFORDSHIRE LIGHT INFANTRY (TERRITORIAL) MUSEUM
Address: T.A. Centre, Harold Street, Hereford.
Open: Monday to Friday 0900 to 1700 (and when Centre is open for training evenings and weekends).

KENT

THE ROYAL DRAGOONS (1ST DRAGOONS)
Final location of museum not yet known.

MUSEUM OF THE CORPS OF ROYAL ENGINEERS
Address: Brompton Barracks, Chatham.
Open: Monday to Friday 1000 to 1230, 1400 to 1630 (excluding public holidays).

THE BUFFS MUSEUM
Address: Stour Street, Canterbury.
Open: Monday to Saturday (April to October) 1000 to 1300, 1400 to 1700; Monday to Saturday (November to March) 1300 to 1600.

THE QUEEN'S REGIMENTAL MUSEUM
Address: Howe Barracks, Canterbury.
Open: Monday to Friday 1000 to 1200, 1400 to 1600; otherwise by appointment.

THE QUEEN'S OWN ROYAL WEST KENT REGIMENT MUSEUM
Address: The Maidstone Museum and Art Gallery, St Faith's Street, Maidstone.
Open: Monday to Saturday, 1000 to 1200, 1400 to 1700.

KENT AND COUNTY OF LONDON (SHARPSHOOTERS) MUSEUM
Temporarily in store, to be rehoused later.

LANCASHIRE

THE REGIMENTAL MUSEUM, THE LOYAL REGIMENT (NORTH LANCASHIRE)

Address: Fulwood Barracks, Preston.

Open: Monday to Friday 0930 to 1230, 1400 to 1630; Saturday and Sunday by appointment.

THE KING'S OWN REGIMENTAL MUSEUM

Address: City Museum, Old Town Hall, Market Square, Lancaster.

Open: Monday to Saturday 1000 to 1730.

REGIMENTAL MUSEUM, 14TH/20TH KING'S HUSSARS

Starting to build up a museum in conjunction with The Manchester Regiment in Manchester.

EAST LANCASHIRE REGIMENTAL MUSEUM

Address: Townley Hall, Burnley (one room).

Open: Monday to Saturday 1000 to 1730; Sunday 1400 to 1700.

LANCASTRIAN BRIGADE MUSEUM

Address: Fulwood Barracks, Preston.

Open: Monday to Friday 0830 to 1230, 1330 to 1645; Saturday 0830 to 1230.

REGIMENTAL MUSEUM, THE SOUTH LANCASHIRE REGIMENT (PWV)

Address: Peninsula Barracks, Warrington.

Open: Monday to Friday 0930 to 1230, 1400 to 1630; weekends by appointment.

EAST LANCASHIRE REGIMENTAL MUSEUM

Address: Fulwood Barracks, Preston.

Open: Monday to Friday 0830 to 1230, 1330 to 1645; Saturday 0830 to 1230.

THE KING'S REGIMENT (LIVERPOOL) MUSEUM
Address: City of Liverpool Museum, William Brown Street, Liverpool 3.

REGIMENTAL MUSEUM, XX THE LANCASHIRE FUSILIERS
Address: Wellington Barracks, Bury.
Open: Monday to Friday 0915 to 1700; Saturday (except November to March) 0915 to 1200.

LEICESTERSHIRE

REGIMENTAL MUSEUM, 9TH/12TH ROYAL LANCERS
It is hoped that a museum will eventually be set up in Leicester.

ROYAL LEICESTERSHIRE REGIMENTAL MUSEUM
Address: Magazine Tower, Leicester.
Open: Daily 1000 to 1800.

LINCOLNSHIRE

17TH/21ST LANCERS MUSEUM
Address: Belvoir Castle, near Grantham.
Open: Wednesday, Thursday, Saturday 1200 to 1800; Sunday (October only) 1400 to 1800; Good Friday, bank holidays (Monday and Tuesday) 1100 to 1900. Open from March to October.

MUSEUM OF THE ROYAL LINCOLNSHIRE REGIMENT
Address: The Keep, Sobraon Barracks, Burton Road, Lincoln.
Open: Monday to Thursday 0900 to 1300, 1400 to 1730; Friday 0900 to 1300, 1400 to 1700.

NORFOLK

THE ROYAL NORFOLK REGIMENT MUSEUM
Address: Britannia Barracks, Norwich.
Open: Monday to Friday 0900 to 1630.

NORTHAMPTONSHIRE

ROYAL PIONEER CORPS MUSEUM

Address: Simpson Barracks, Wootton, Northampton.

Open: Not yet known.

This museum is in formative stages.

THE MUSEUM OF THE NORTHAMPTONSHIRE REGIMENT

Address: Gibraltar Barracks, Barrack Road, Northampton.

Open: Tuesday to Saturday 0930 to 1230, 1400 to 1630.

NORTHUMBERLAND

REGIMENTAL MUSEUM, THE KING'S OWN SCOTTISH BORDERERS

Address: The Barracks, Berwick-upon-Tweed.

Open: Monday to Friday 0900 to 1200, 1300 to 1630; Saturday 0900 to 1200; other times on application to Curator.

ROYAL NORTHUMBERLAND FUSILIERS REGIMENTAL MUSEUM

Address: The Armoury, Fenham Barracks, Newcastle-upon-Tyne 2.

Open: Monday to Friday 1000 to 1600; or by appointment.

REGIMENTAL MUSEUM, 15TH/19TH THE KING'S ROYAL HUSSARS

Not yet established.

NORTHUMBERLAND HUSSARS MUSEUM

Not yet established.

NOTTINGHAMSHIRE

THE SHERWOOD FORESTERS (NOTTINGHAMSHIRE AND DERBYSHIRE) MUSEUM

Address: The Castle, Nottingham.

Open: Daily 1030 to dusk.

SHROPSHIRE

THE KING'S SHROPSHIRE LIGHT INFANTRY AND THE HEREFORDSHIRE LIGHT INFANTRY MUSEUM

Address: Sir John Moore Barracks, Copthorne, Shrewsbury.

Open: Monday to Friday 1000 to 1200, 1500 to 1600.

REGIMENTAL MUSEUM, 1ST THE QUEEN'S DRAGOON GUARDS

Address: Clive House, Shrewsbury.

THE SHROPSHIRE YEOMANRY REGIMENTAL MUSEUM

Address: Territorial House, Sundorne Road, Shrewsbury.

Open: Monday to Saturday 0900 to 1700.

SHROPSHIRE YEOMANRY MUSEUM

Address: Territorial House, Sundorne Road, Shrewsbury.

Open: Monday to Friday 0900 to 1700 (and when House is open for training evenings and weekends).

(This museum also includes the Shropshire R.H.A. Museum.)

KING'S SHROPSHIRE LIGHT INFANTRY (TERRITORIAL) MUSEUM

Address: The Drill Hall, Coleham, Shrewsbury.

Open: Monday to Friday 0900 to 1700 (and when Drill Hall open for training evenings and weekends).

SOMERSET

THE SOMERSET LIGHT INFANTRY MUSEUM

Address: 14 Mount Street, Taunton.

Open: Monday to Friday 0900 to 1200, 1400 to 1700; Saturday 0900 to 1200.

STAFFORDSHIRE

THE STAFFORDSHIRE REGIMENTAL MUSEUM

Address: Whittington Barracks, Lichfield.

Open: Monday to Friday 1000 to 1630; weekends and bank holidays by arrangement with Curator.

SUFFOLK

SUFFOLK REGIMENT MUSEUM
Address: The Keep, Gibraltar Barracks, Bury St Edmunds.

SURREY

NATIONAL ARMY MUSEUM
Address: Royal Military Academy, Sandhurst, Camberley.
Open: Monday to Saturday 1000 to 1700; Sunday 1100 to 1700. Admission free.

R.A.O.C. MUSEUM
Address: R.A.O.C. Training Centre, Deepcut, Camberley.
Open: Monday to Friday 1000 to 1200, 1400 to 1600 (public holidays excepted).

REGIMENTAL MUSEUM, THE QUEEN'S ROYAL SURREY REGIMENT
Address: Surbiton Road, Kingston-upon-Thames.
Open: Monday to Friday (except public holidays) 0930 to 1230, 1330 to 1600; Sunday 1000 to 1200 (by arrangement).

SUSSEX

ROYAL MILITARY POLICE MUSEUM
Address; Roussillon Barracks, Chichester.
Open: Monday to Friday 0900 to 1800; weekends and public holidays by appointment only.

THE ROYAL SUSSEX REGIMENT MUSEUM
Address: Chichester City Museum, 29 Little London, Chichester.
Open: Monday to Saturday (April to September) 1000 to 1800; Tuesday to Saturday (October to March) 1000 to 1700.

WARWICKSHIRE

THE ROYAL WARWICKSHIRE REGIMENTAL MUSEUM

Address: Regimental Headquarters, The Royal Warwickshire Fusiliers, St John's House, Warwick.

Open: Monday, Wednesday, Thursday, Friday 1000 to 1230, 1400 to 1630; Saturday 1430 to 1700; Sunday (May to September) 1430 to 1700.

THE QUEEN'S OWN HUSSARS MUSEUM

Address: The Lord Leycester Hospital, High Street, Warwick.

Open: Monday to Saturday (April to October) 1000 to 1800. Monday to Saturday (November to March) 1000 to 1630.

WARWICKSHIRE AND WORCESTERSHIRE YEOMANRY MUSEUM

Address: Drill Hall, Priory Road, Warwick.

Open: Monday to Friday 0900 to 1630; Saturday and Sunday by appointment.

WORCESTERSHIRE

THE WORCESTERSHIRE REGIMENT MUSEUM

Address: Norton Barracks, Worcester.

Open: Monday to Friday 0900 to 1230, 1400 to 1600.

YORKSHIRE

THE GREEN HOWARDS MUSEUM

Address: Gallowgate, Richmond.

Open: Monday to Saturday (15 April to 1 November) 1000 to 1700; Sunday (15 April to 1 November) 1400 to 1630.

THE WEST YORKSHIRE REGIMENT (14TH FOOT) AND P.W.O. MUSEUM

Address: Imphal Barracks, York.

Open: Monday to Friday 0900 to 1230, 1330 to 1630 (except public holidays).

THE DUKE OF WELLINGTON'S REGIMENTAL MUSEUM

Address: Bankfield Museum, Boothtown Road, Halifax.

Open: Monday to Saturday (April to September) 1100 to 1900; Monday to Saturday (October to March) 1100 to 1700; Sunday 1430 to 1700.

4TH/7TH ROYAL DRAGOON GUARDS MUSEUM

Address: Bankfield Museum, Boothtown Road, Halifax.

Open: Monday to Saturday (April to September) 1100 to 1900; Monday to Saturday (October to March) 1100 to 1700; Sunday 1430 to 1700.

CASTLE MUSEUM

Address: York.

Open: Monday to Saturday (April to September) 0930 to 1930; Sunday (April to September) 1400 to 1930; Monday to Saturday (October to March) 0930 to 1630; Sunday (October to March) 1400 to 1630. Admission 1*s.* 6*d.*

REGIMENTAL MUSEUM, THE YORK AND LANCASTER REGIMENT

Address: Endcliffe Hall, Endcliffe Vale Road, Sheffield 10.

Open: Monday to Friday 0900 to 1630; Saturday and Sunday by appointment only.

K.O.Y.L.I. MUSEUM

Address: Regimental Headquarters, The King's Own Yorkshire Light Infantry, Wakefield Road, Pontefract.

Open: Monday to Friday 0900 to 1700.

THE EAST YORKSHIRE REGIMENT (15TH FOOT) MUSEUM

Address: 11 Butcher Row, Beverley.

Open: Wednesday, Thursday and Friday 1400 to 1600 (except public holidays).

SCOTLAND

SCOTTISH UNITED SERVICES MUSEUM

Address: Crown Square, The Castle, Edinburgh 1.

Open: Monday to Saturday (summer) 0930 to 1800; Sunday (summer) 1100 to 1800; Monday to Saturday (winter) 0930 to 1630.

QUEEN'S OWN HIGHLANDERS (SEAFORTH AND CAMERONS) MUSEUM

Address: Fort George, Inverness-shire.

Open: Monday to Saturday (April to September) 1000 to 1900; Sunday (April to September) 1400 to 1600; Monday to Saturday (October to March) 1000 to 1600.

THE ARGYLL AND SUTHERLAND HIGHLANDERS REGIMENTAL MUSEUM

Address: The Castle, Stirling.

Open: Sunday to Saturday (May to September) 1000 to 1800; Monday to Friday (October to April) 1000 to 1600.

REGIMENTAL MUSEUM, THE CAMERONIANS (SCOTTISH RIFLES)

Address: 129 Muir Street, Hamilton.

Open: Monday to Friday 0900 to 1630.

THE BLACK WATCH MUSEUM

Address: Balhousie Castle, Perth.

Open: Monday to Friday (1 May to 30 September) 1000 to 1200, 1400 to 1700; Sunday (1 May to 30 September) 1400 to 1700; Monday to Friday (1 October to 30 April) 1000 to 1200, 1400 to 1600; Sunday (October and April only) 1400 to 1700. Saturday by special arrangement, parties of 12 or more, 24 hours' notice.

THE ROYAL SCOTS GREYS

Address: The Royal Scots Greys' Room, Scottish United Services Museum, The Castle, Edinburgh 1.

Open: Monday to Saturday (summer) 0930 to 1800; Sunday (summer) 1100 to 1800; Monday to Saturday (winter) 0930 to 1630.

Address: Home Headquarters, The Royal Scots Greys, The Castle, Edinburgh 1.

Open: Monday to Friday 0900 to 1700.

Address: The Binns, Linlithgow (home of the Dalyell family, in National Trust premises, and has relics of General Tam Dalyell, who raised the Regiment, in the grounds).

Open: Sunday to Saturday (summer) 1030 to 1800; Saturday and Sunday (winter) 1400 to 1600.

THE ROYAL SCOTS REGIMENTAL MUSEUM

Address: Regimental Headquarters, The Royal Scots, The Castle, Edinburgh 1.

Open: Sunday to Saturday (June to September) 0930 to 1800; Monday to Friday (October to May) 0930 to 1600.

AYRSHIRE YEOMANRY MUSEUM

Address: Yeomanry House, Ayr.

Open: As required.

GORDON HIGHLANDERS MUSEUM

Address: Viewfield Road, Aberdeen.

Open: Wednesdays and Sundays, 1400 to 1700. Admission 1*s*, Children 6*d*.

THE ROYAL HIGHLAND FUSILIERS MUSEUM

Address: 518 Sauchiehall Street, Glasgow C.2.

Open: Monday to Friday 0900 to 1230, 1400 to 1700.

THE SCOTTISH HORSE MUSEUM

Address: The Cross, Dunkeld, Perthshire.

Open: Sunday to Saturday (from Easter to the end of October) 1400 to 1600, 1800 to 1930.

Lowland Brigade Depot Museum

Address: Glencorse Barracks, Milton Bridge, Penicuik, Midlothian.

Open: Monday to Saturday 0900 to 1230, 1400 to 1630.

WALES

Regimental Museum, the South Wales Borderers

Address: The Barracks, Brecon.

Open: Sunday to Saturday, 0900 to 1700 (including public holidays).

Regimental Museum, the Welch Regiment

Address: The Barracks, Whitchurch Road, Cardiff.

Open: Monday to Friday 1000 to 1200, 1400 to 1600.

The Royal Welch Fusiliers Regimental Museum

Address: The Queen's Tower, Caernarvon Castle, Caernarvon.

Open: Daily (June to September) 0930 to 1930; Monday to Saturday (March, April, October) 1930 to 1730; Sunday (March, April, October) 1400 to 1730; Monday to Saturday (November to February) 0930 to 1600; Sunday (November to February) 1400 to 1600. (Corresponding to Castle opening times.)

NORTHERN IRELAND

Regimental Museum, the Royal Ulster Rifles

Address: 5 Waring Street, Belfast.

Open: Monday to Friday 1000 to 1600; Saturday 0930 to 1200; parties by special appointment only.

Regimental Museum, the Royal Irish Fusiliers

Address: Sovereign's House, The Mall, Armagh.

Open: Monday to Friday 1000 to 1230, 1400 to 1630; Saturdays and Sundays by prior arrangement.

REGIMENTAL MUSEUM, 5TH ROYAL INNISKILLING DRAGOON GUARDS
NORTH IRISH HORSE MUSEUM
A museum is to be established in Carrickfergus Castle, Carrickfergus, Co Antrim.

REGIMENTAL MUSEUM, THE ROYAL INNISKILLING FUSILIERS
Address: St Lucia Barracks, Omagh, Co. Tyrone.
Open: By appointment only.

REGIMENTAL MUSEUM, THE QUEEN'S ROYAL IRISH HUSSARS
A museum is to be established in Carrickfergus Castle, Carrickfergus, Co Antrim.

INDEX

References to illustrations are in italic type

Index

Index

Index

Index

Index